A book fr

All
http://north-s

Pierre Loti
(1850-1923)

of the French Academy

———————

Aziyadé

1879

Translated by
William Needham

TAKEN FROM THE NOTES AND LETTERS
OF A BRITISH NAVAL LIEUTENANT WHO
SERVED WITH THE TURKISH FORCES
FROM 10 TH MAY 1876 AND WAS KILLED
WITHIN THE WALLS OF KARS 27 TH OCTOBER 1877.

PARIS

———

CONTENTS.

PREFACE BY PLUMKETT,

A FRIEND OF LOTI

In every well constructed romance, a description of the hero is obligatory. But this book is not a romance, or, at least, is not one that has been better managed than the life of its hero. Besides, describing Loti, whom we may indeed be very fond of, is not without its difficulties and might well baffle even the cleverest.

To form a physical image of him, dear reader, go to Musset: open 'Namouna, a tale from the Orient' and read:

> 'Shoulders squared; freshly groomed; ....
> Hands of a patrician, demeanour proud and tense;
> Finest among all his features were his eyes.'

Like Hassan he was very cheerful, yet could be very sullen; shamefully credulous, yet could be utterly indifferent. Be it for good or ill he was always one to persist; but we preferred him to the self-centred Hassan, and he might be said to have borne more resemblance to Rolla ...'

> 'In more than one soul can two things be seen together :
> 'The sky, colouring the almost still waters.
> And the mud,--dreary seabed, horrid, gloomy and idle.'

(VICTOR HUGO, 'Les Ondines' .)

PLUMKETT.

BOOK ONE - SALONIKA

I

16 May 1876.

... A beautiful day in May, lovely sunshine, clear sky ... When the foreign boats arrived, the hangmen on the quayside put the finishing touches to their work: six dangling bodies performed their horrible final contorsion before the crowd ... Windows, roofs were completely filled with spectators; on a balcony nearby, Turkish officials smiled at the familiar spectacle.

The Sultan's government had gone to no great expense with the instrument of torture; the gallows were so low that the bare feet of the condemned men touched the ground. Their toenails, clenched in a snarl, raked the sand.

II

The execution over, the soldiers withdrew, and until nightfall the dead remained on public view. All afternoon in the delightful Turkish sun the six cadavers, upright on their feet, displayed their hideous grimaces, surrounded by unconcerned passers-by and by groups of young women standing in silence.

III

The governments of France and Germany had demanded the executions be combined, as redress for the massacre of the consuls which had attracted much attention in Europe at the start of the near-eastern crisis.

All European nations had sent huge battleships to Salonika. Britain had been one of the first to do so, hence my arrival here on board one of Her Majesty's corvettes.

IV

One fine spring day, among the first during the period when the Macedonians--not long after the massacres and three days after the hangings--granted us permission to go into Salonika, I found myself around four o'clock in the afternoon pausing in front of the closed door of an old mosque to watch two storks fighting.

I can still picture the scene in that street in the old Moslem quarter. Abandoned houses rose on either side of the narrow, winding alleyways;

and projecting halfway across them were '*shaknisirs*', a weird kind of observatory, where large balconies had been closed off with iron grills from which, through small invisible holes, one could watch those going by below.

Oat shoots sprouted between the black cobblestones, and branches, covered in greenery, spread along the roofs. One had chance glimpses of

a clear blue sky; and every breath caught the mildness of the air and the wonderful smell of Maytime.

The people of Salonika still preserved an attitude towards us of constraint and hostility; moreover, the order was, that at all times in the street we were to carry our swords and all other weapons 'at the trail'. Every now and then turbaned figures would go by, following the wall; and not a single female head was to be seen behind the discreet grills of the women's appartments, the '*haremlikes*'. A dead city, one might have said.

I thought I was perfectly alone; then I experienced a strange feeling, and realised that close to me, from behind thick iron bars at head height, two large green eyes were staring into mine.

The eyebrows were brown, and the slight frown brought them so close that they met; the gaze suggested a combination of vigour and openness; it contained so much freshness and youth it could have been taken for a child's.

The young woman, whose eyes these were, rose, and from her waist upward one could tell she was wrapped in the long, stiff folds of a Turkish-style cape, a '*feredge*', made of green silk, embroidered with silver. A white veil carefully enveloped her head, revealing only her forehead and her large eyes. The pupils were indeed green: that shade of sea green, which was celebrated in the past by the poets of the Orient.

The young woman was Aziyade.

V

Aziyade stared at me. In front of a Turk she would have hidden her face; but I was a 'giaour', an infidel, which doesn't count as a man: at the very most he's a curiosity to be observed at leisure. She appeared surprised that one of the foreigners, who had come, with such terrible war machines, to threaten her country with such terrible war machines, could be a very young man whose appearance aroused in her neither repugnance nor fear.

VI

When I returned to the quay all the boats belonging to the squadrons had left. I'd been a trifle captivated by those green eyes even though I hadn't yet discovered the exquisite face hidden by white veil; three times I'd walked past the mosque that had the storks, and time had gone by without my noticing.

As if I'd summoned them, the obstacles between this young woman and me came thick and fast: the impossibility of exchanging a single thought with her, or of speaking to her, or of writing to her. Going ashore after six in the evening was forbidden; and at all times one had to be armed; likely departure within a week, and never to return; and above all, the fierce surveillance that was maintained over the harems.

I watched the last British ship draw further and further away; the sun would very soon disappear, and, not having come to any decision, I sat myself down beneath the awning of a Turkish café.

VII

Straightaway a bunch of men gathered round me; these were the boatmen

and porters who slept on the quayside under the stars; they wanted to know why I'd stayed behind on shore; they were waiting there in the hope that I might perhaps have need of their services.

Among this group of Macedonians I noticed a young man with a very quaint type of beard, that was divided up into little curls like the ones you see on the most ancient of the statues in that country. He was sitting on the ground in front of me, inspecting me with great curiosity; my clothes and, most of all, my boots appeared to interest him keenly. He stretched himself in the cuddly way a large angora cat would, and when he yawned he displayed two rows of tiny teeth that shone like pearls. His head was very fine and there was a great gentleness in his eyes that glowed with honesty and intelligence. He was dressed only in rags, his legs were bare, his shirt in tatters, yet he was as clean as a cat.

This young fellow was Samuel.

VIII

These two beings, whom I met on the very same day, were soon to become part of my life, and would, over the next three months, risk their lives on my account. If anyone had told me that then, I would have been greatly surprised. Later they would leave their own country behind and follow me to Istanbul, where we were destined to spend winter under the same roof.

IX

Samuel screwed up enough courage to use the tiny amount of English words he knew, and said:

'Do you want to go on board?'

He continued in Sabir:

'Te portarem col la mia barca.' (I'll take you on my boat).

Samuel could speak Sabir, then; I saw straightaway what an advantage an intelligent and determined lad might prove to be for the hare-brained venture that was going round in my head and just beginning to take some sort of shape.

Money was one means whereby I could enlist the services of this beggarly youth, but I'd very little. Besides, he was bound to be respectable and, being so, would definitely not allow himself to act as go-between for a young man and a young woman.

X

'To WILLIAM BROWN,

Lieutenant with the 3rd Line Infantry Regiment,

Salonika, Jun 2th.

'In the beginning it was nothing more than an intoxication of my imagination and senses; later something new came along: love, or something very close to it; I was surprised and enchanted.

'Had you been able today to follow your friend Loti through the streets of an old and forgotten part of the city, you would have seen him climb the steps to a house that looks as if it belongs to a fantasy world. The door closes behind him mysteriously. He chooses the section where they can change the scenes; the backgrounds are familiar to him. Before, you'll remember, it was to enjoy the favours of Isabelle B..., the star: the scene was in a hansom cab or Hay Market Street, or where the mistress of the famous Martyn lived; this business of changing backgrounds goes back a long time, and it would take more than Oriental dress to add a touch of novelty to it.)

'The melodrama begins. First setting: a dingy old room that looks

rather poor, though it has plenty of Oriental colour. 'Hookahs' lie around the floor along with weapons.

'Your friend Loti is standing in the centre and three old Jewesses are busying themselves wordlessly around him. Their garb is picturesque, and their noses hooked. They wear long, jackets trimmed with spangles, necklaces of strung sequins; their hair is tied back with green silk ribbons. They briskly relieve him of his officer's uniform and begin to dress him Turkish style, kneeling so they can start with the gilt leggings and the garters. Loti maintains the look of sombre concern, appropriate for the hero of a musical drama.

'The three old women put into his belt one or two daggers whose silver hilts are inlaid with coral and their blades damascened with gold; they slip over onto his shoulders a gold embroidered jacket with floating sleeves and on his head they place a turban, a 'tarboosh'. All done, they tell Loti by gesture how handsome he looks, and bring a large mirror.

'Loti finds he's not at all displeased with the image and smiles sadly at this altered appearance that could get him killed; and then he disappears by a back door and makes his way across an entire crazy city of oriental bazaars and mosques; he passes undetected through the multi-coloured throngs that are dressed in the brilliant hues beloved by the Turks. Occasionally one of the white veiled women would simply remark: 'See that well dressed Albanian and his beautiful weapons.'

'Following your friend Loti any further, my dear William, would be inadvisable. My goal will be reached when I'm in the arms of a Turkish woman who is married to a Turkish man--a crazy enterprise at any time, and, things being as they are, one that cannot be spoken of.-- With her Loti will spend an hour of total rapture, and risk her head and the heads of several others, as well as all manner of diplomatic complications.

'You will say that there has to be a deep-seated egoism at work to reach this stage. I shan't contradict you; but I'm now of the view that to do that which pleases me is a good thing, and that one has to do one's best to add spice to life's utterly tasteless dish.

'You have nothing to reproach me for, dear William: I've written you a good, long letter. I have no more confidence in your affection than I have in anyone else's; but you are one of a number of people I've met here and there as I've travelled the world, one of those whom its been a pleasure to be with and to share my impressions. If there's anything in my letter that's a touch unrestrained I cannot be blamed: I've been drinking Cyprus wine.

'It has now worn off. I went up onto the bridge to breathe some sharp night air. Salonika looked a wretched place. Its minarets seemed like a lot of old candles placed on top of a town that was dark and dirty, the sort of town where the vices of Sodom are rife. When the damp air hits me like an ice-cold shower and nature puts on a dull, pitiful look, I go back to being my normal self where I find only life's sickening emptiness and immense boredom.

'I'm thinking of going to Jerusalem soon. There I shall try to pull together what few scraps of faith I possess. For the moment my religious and philosophical beliefs, my moral principles, my social

theories, etc, can all be depicted by that grand personality, the policeman.

'I'll probably be coming over to Yorkshire in autumn and will see you. Meanwhile, I take my leave of you, and remain your devoted friend.

LOTI.'

XI

One of the most troubled periods in my life were the closing days of May 1876. I had gone through much pain with the result that, for a long time afterwards I had felt crushed, listless, my heart empty; but all of that went away, and it was the force of youth that brought about my new awakening. Suddenly I found myself alone in life; the last of my beliefs were gone, which meant I had free rein to do whatever I chose. From the ashes of my former self something like love was beginning to take shape, and over this renaissance of mine the Orient was casting its powerful spell and troubling my senses.

XII

She had come, together with the three other wives of her master, to live in a '*yali*', a country villa, on the road to Monastir; here she would be under lighter supervision.

I used to arrive in the daytime, armed. When there was a heavy sea, a cutter would drop me at the docks among the crowd of boatmen and fishermen; and to Samuel, who would just happen to be where we could see eachother, I would signal my orders for the night.

I spent many a day wandering along the Monastir road. The land about was bare and depressing; as far as the eye could see were ancient cemeteries; ruins of marble tombs, their mysterious inscriptions eaten away by lichen; whole stretches where stood granite menhirs; Greek, Byzantine, and Moslem graves covered the ages-old earth of Macedonia where great nations of the past had left their dust. Every now and then appeared the sharp silhouette of a cypress or a huge plane tree giving shade to Albanian shepherds and their goats; on the parched ground broad flowers, pale lilac in colour, spread their sweet smell of honeysuckle beneath an already scorching sun. The slightest detail of that landscape is stored in my memory.

The soft warmth of the night brought with it an unvarying stillness;

sounds from cicadas uniting with the silence; air that was pure yet filled with the perfumes of summer; the sea motionless; the sky as bright as bright as any I had known in the tropics.

She wasn't mine, not yet; but the only barriers between us were material ones: her master and the iron gratings at her windows.

I spent my nights waiting for her, waiting for that moment which was sometimes very short, when I could touch her arms through those grim bars and in the darkness kiss her white hands with their oriental rings.

Finally, at a certain time in the morning before daybreak, in accordance with the plan I had agreed with the officers on watch, I would return to my corvette.

XIII

My evenings were spent with Samuel. I saw some strange things with him in the boatmen's taverns; and in beggars' courtyards and in thieves' bars run by Jews from Turkey I was able to study at first hand behaviour that few people could have witnessed.

The clothes I wore in these hell-holes were the sort of clothes Turkish sailors wore touring the Salonika harbourside at night. Samuel presented an odd contrast in such surroundings; his lovely, gentle face shone out all the more against the foil of all the murk. Bit by bit I was getting to be fond of him and his refusal to be of any use to me with regard to Aziyade made me think all the more highly of him.

All the same, I saw some weird sights at night in the company of this little vagabond: there was a certain bizarre kind of prostitution that went on down in the wine cellars where customers drank themselves stupid on 'mastic' and 'raki'...

XIV

One balmy June night the two of us were lying out on the ground waiting for two o'clock, the time agreed, to come. I can remember that beautiful starry night when all one could hear was the faint sound of a tranquil sea. Patterning the mountainside were the black, peardrop shapes of the cypresses and the huge dark masses of the plane-trees. Here and there centuries-old boundary stones marked the forgotten domain of some long gone Dervish. The dry grass, the moss, and the lichen all smelled good; it was good too to be outside in the fields on such a night; it felt good to be alive.

But Samuel appeared to be performing his nightly errand with dreadful disdain and ceased replying to me. So, I took hold of his hand, as friends do, and, in Spanish, said something along these lines: 'My dear Samuel, every night you sleep on the hard ground or on planks; it's better here on this grass where you have the lovely smell of the wild-thyme as well. Go to sleep, and you'll feel better afterwards. Aren't you happy being with me? What have I done to upset you?'

His hand trembled in mine and he was gripping it with unnecessary force.

'Che volete,' he said in a voice that was deeply perturbed, 'Che volete mi? (What do you want from me?)'

For a moment something unspeakably vile had come into poor Samuel's head:--in the older parts of the Orient all things are possible. Then he covered his face with his arms, and stayed where he was, terrified at himself, transfixed, and trembling.

But ever since that strange episode he has devoted himself, body and soul, to my service. Every night he risks his liberty and his life by entering the house where Aziyadé lives. When he goes in the darkness to fetch her he has to cross a cemetery that for him harbours ghostly apparitions and mortal terrors. He rows around in his little boat until daybreak in order to keep watch for the larger vessel, or he waits for me all night on the fifth flagstone on the quayside at Salonika with the bodies of fifty sleeping vagrants strewn about for company.

It's as if his own personality has been absorbed into mine and I feel it accompanying me everywhere I go, wherever that may be and whatever disguise I may have chosen to put on; he is ready to defend my life with his own

XV

LOTI to PLUMKETT, Lieutenant of Marine

Salonika, May 1876.

'My dear Plumkett,

'You may, without fear of ever boring me, tell me about all the things, dreary or preposterous or even light-hearted that are going through your head right now. Since I always place you outside the common herd, reading what you have to say will never cease to be a pleasure.

'Your letter was handed to me at the end of a dinner accompanied with Spanish wine, and I recall being, at first reading, a wee bit bewildered at the way it had been put together. You're a funny fellow and no mistake, but then I always knew that. You're famous for having a mind of your own. But, I assure you, that's by no means all that I was able to draw from your long letter. I could see that you've had to endure much suffering, and that is something we have in common.

'Ten long years ago, when I was sixteen, I was packed off to London and left to my own devices. I sampled all available pleasures, but, at the same time, I believe there was practically no kind of sorrow that I was spared. I'm feeling terribly old despite my extremely youthful physique, which I owe to my fencing and acrobatics.

'Confidences, of course, serve no purpose; for there to be fellow-feeling between us it is enough that you have suffered.

'I see also that I've been fortunate enough to inspire in you some degree of affection; thank you for that. Ours will be, if you greatly wish it, what you call an 'intellectual friendship', and our relationship will ease our passage through the gloomy periods of life.

'On the fourth page of your letter doubtless you were letting your pen run away from you when you wrote: 'boundless affection and devotion'. If those were your thoughts then you'll clearly see, my dear fellow, that you still have something of your youth and freshness about you, and that they are not completely gone. No one has great experience than I have of such fine friendships-- friends for life, friends until death--but, you see, that's what one has when one is eighteen; at twenty-seven they're all done with and one is only devoted to oneself. It's distressing, what I'm saying, but it's only too true.'

Salonika, June, 1876.

It was a joy, making those early morning runs to Salonika, arriving before the sun had risen. The air was so light and the cool so delicious that life's troubles might not have existed; it was as if one were actually absorbing well-being. Several Turks, some dressed in red robes, others in green or orange, were beginning to move about the arched-over streets, which, as yet, were barely lit by the thin glimmer of dawn light.

When I had Thompson, the ship's engineer, with me, he played the comic opera role of the confidant, and we roamed extensively round the old streets of the city at a time of day when to do so was forbidden to us, and when our uniforms least accorded with regulations.

Evening time presented us with an enchantment of a different kind: everywhere was either pink or gilt. Mount Olympus, aglow as if with burning embers or molten metal, was mirrored on the smooth surface of

the sea. No mist in the air: it seemed that the atmosphere no longer existed and that the mountains--so distinct and precise were their

furthest ridges--were cut-outs with a huge void behind them.

We often sat of an evening among the crowds on the quays looking out across the tranquil bay. Oriental barrel organs accompanied by tiny bells and Turkish jingle poles churned out their weird tunes; public roads were jammed with cafés, their small tables already laid, yet never enough of them to meet the demand for hookahs, Skyros wine, Turkish Delight, and raki.

It made Samuel happy and proud when we invited him to our table. He was always somewhere nearby so that he could signal to me wherever it was I was to meet Aziyade; then I'd tremble with excitement thinking about the night ahead.

XVII

Salonika, July 1876.

Last night Aziyade told Samuel that he was to remain with us. I looked at her in astonishment: she asked me to sit between the two of them, then she started talking to him in Turkish. It was a conversation she was wanting, our first; Samuel was to be our interpreter. For a month we had shared an ecstasy of the senses and it had never been possible to exchange even a single thought and until now we remained as ignorant of each other as strangers.

'Where were you born?... Where did you live?... How old are you?... Do you have a mother?... Do you believe in God?... Have you been in the country of the black men?... Have you had a lot of mistresses?... Are you a lord in your own country?'

She told me she was Circassian and had arrived in Constantinople while still a child, together with another girl the same age as herself. A merchant had sold her to an old Turk who had brought her up intending to give her to his son; the son had died, so had the old Turk; by then she was sixteen and extremely pretty; then this man spotted her in Stamboul and brought her back to his house in Salonika.

'She says,' Samuel translated, 'that her God is not the same as yours, and she's not very sure, from what it says in the Koran, that women have a soul as men do; she thinks that when you're gone from this country, you and she will never see each other again, even when you're dead, and that's what's making her cry.'

'Now,' said Samuel with a laugh, 'she asks if you want to jump into the sea with her right now and drift down to the bottom in each other's arms, and then I'll take the boat back and say that I've not seen either of you.'

'With pleasure,' I said, 'provided she stops crying. Let's do it rightaway, then it'll be finished with.'

Aziyade understood. Trembling she slid her arms around my neck and we both leaned out over the water.

'Don't do that!' shouted a frightened Samuel, restraining us in an iron grip, 'Your last kiss would be vile. People who drown together bite eachother's face and they die with horrible grimaces.'

He had spoken in Sabir with a violent crudeness which doesn't translate.

It was time for Aziyade to set off back: and a momemt later, she left us.

XVIII

'My dear Loti,

'I have a vague recollection of sending you a letter last month that

you wouldn't have made head or tail of, neither rhyme nor reason. One of those letters where one's first thoughts take command or where one's imagination gallops ahead and one's pen just trots along, often stumbling like some old hack.

'Letters like those are never read over before they're sealed, otherwise they'd never be sent. Some digressions, more or less pedantic

and unclear, followed by some nonsensical items that would have been

unworthy of the 'Tintamarre'. Then, for the final flourish, an auto-panegyric from a misunderstood individual, affecting self-deprecation in hopes of harvesting compliments, such as those you have been kind enough to send. In short, the whole thing: quite ridiculous!

'And those protestations of devotion!--Ah! that set the old hack at a gallop. You responded to this part of my letter as would that writer have done, who lived sixteen centuries before our era: he had tried everything—being a great king, a great philosopher, a great architect, keeping six hundred wives, etc.--but grew so bored and weary with it all that in his old age, his reflections complete, he declared that everything was nothing but vanity. For all its Ecclesiastes style, your reply brought me nothing I did not know already. I am very much of the same mind as you regarding all things and everything, so much so that I question whether I'll ever be able to discuss a topic with you other than in the same way Pandore conversed with his sergeant. In the moral category we have absolutely nothing to teach eachother.

"Confidences,' you tell me, 'serve no purpose.' I admire your insight even more: I like to have an overall view of persons and things; I like to discern their principal features. As for details, I've always had a perfect horror of them.

"Boundless affection and devotion'! As you wish! It was one of those warm impulses, one of those felicitous glimpses that privileges one to see one's better self. Be in no doubt that the words are sincere. And if they are no more than flashes, who is to blame for that? Isn't it true that you and I are in no way responsible for the flaw in our nature? Is it, then, he who created us, only to leave us unfinished, sensitive to the noblest aspirations, but lacking an ability proportionate to our concepts? Or is there no one at all who is to blame? Better to leave all this doubting, I think, and move on to something else.

'Thank you for your comments regarding my candidness about my feelings—though I don't believe a word of what you say. They've seen too much service, or rather, I've called upon their service too often, for them not to have faded from use. I could say my feelings are reserved for occasions--talking of which, let me remind you just how often right

occasions come along. I'll have you note also that

there are things that gain solidity the more usage rubs away the gleam of their newness; an example I can give you from the noble profession we both follow would be old cordage.

'Naturally, I am very fond of you. And we needn't refer to it again. I declare, once and for all time, that you are indeed very gifted and it would be a pity if you let the better part of you exhaust itself doing acrobatics. Now that's off my chest, I will cease boring you to death with talk of affection and admiration and give you a few personal details of my own.

'Physically I'm in good shape; morally I'm undergoing treatment which

consists in preventing my brain turning itself inside out, and in

fitting a regulator to my sensibility. In this world everything hinges

on balance--within us and without. If sensibility gets the upper hand, it always does so at the expense of reason. The more you are a poet, the less you are a land-surveyor; and, in life, a small amount of geometry is called for; worse still, a great deal of arithmetic. I believe, God forgive me, that what I have just written comes very close to being common sense!

'Ever yours,

'PLUNKETT'

XIX

Night of July 27. Salonika.

At nine o'clock the officers on board, one after the other, return to their quarters; as they go they wish me luck and good night: they all know my secret.

All the while I'm looking anxiously across at the sky over old Olympus, whence, too often for my liking, come big, coppery clouds indicating

thunderstorms and torrential rain.

As yet, all is clear over there and the summit of that myth-making mountain stands out in bold relief against the deep blue of the night sky.

I go below to my cabin, change my clothes, and come back on deck.

Now, as on every night, the anxious waiting: one hour, two hours go by, the minutes drag and seem as long as nights.

At eleven o'clock, a faint sound of oars on the calm sea; a tiny, far off speck comes nearer, slipping across the water like a shadow. It's Samuel's row boat. The guards on duty raise their rifles, hold him in their sights and call out to him. Samuel doesn't answer; then the rifles are lowered--the guards have been given secret instructions concerning him. Within moments he draws alongside.

Merely for show they hand him my nets and other tackle; I jump into the small boat and we move off. I remove the cloak covering my Turkish attire, and the transformation is complete. My gold-embroidered jacket glimmers slightly in the darkness. The breeze is soft and warm, and Samuel pulls soundlessly in the direction of the shore.

Another boat is already there waiting. On it are a hideous old negress

with a blue cloth wrapped round her, an old Albanian manservant, dressed picturesquely and armed to the teeth, and a woman hidden behind such a quantity of veil that all one sees is a mass of white.

Not a word is said as the two servants transfer to Samuel's row boat, which then departs.

I am left alone with the veiled woman, who, like a white phantom, remains still and silent. My hands are already on the oars and we move away, heading in the opposite direction, out to sea. My eyes are fixed on her and I wait anxiously for her to move or make a sign.

When she feels we are far enough out, she holds out her arms to me; I have been waiting for this sign beckoning me to her side. I tremble as we touch and am pierced by an overwhelming langour; her veil is infused with oriental perfumes, her skin feels firm and cool.

I have loved another woman more, one I no longer have the right to see again; but never before have my senses known such intoxication.

XX

Aziyade's boat contains silk mats, cushions, and Turkish coverings.

Casual touches of oriental refinement are everywhere, and the whole seems more like a floating bed.

What a curious situation is ours: we are prevented from exchanging a solitary word. All manner of perils are gathered round this bed of ours, drifting above the deep; we look like beings who have met with the sole aim of enjoying together the heady charms of the impossible.

In three hours time we will have to leave. That will be when the Great Bear has turned itself around in the immense sky. Each night we follow its unvarying course —this pointer on a clockface counting our hours of ecstasy when we are oblivious of the world and of life, and when the first kiss lasts until dawn. I could liken it to being in an African desert and having a burning thirst that the first draught of cool water

cannot satisfy but only make one want more.

At one o'clock the silence of the night is shattered by a rowdy din. We

hear the sounds of harps and of women's voices, and then come warning shouts which we have barely enough time to heed. A cutter from the 'Maria Pia', sweeps past; their debauch involves mainly Italian officers, most of them drunk;--they only just miss sinking us.

By the time we rejoined Samuel's boat the Great Bear had progressed beyond its highest point, and we could hear cocks crowing in the distance.

Samuel, wrapped in my cloak, was asleep aft in the well of the boat; the negress, crouched up like a monkey, was asleep in the bows; between them the old Albanian slept, bent over his oars.

The two servants were reunited with their mistress, and their boat with all three aboard moved silently away. Steadily my eyes followed the white form of the young woman who was lying, completely motionless, exactly where I had left her, still warm from my kisses, her skin moistened by the dew.

The three o'clock bells sounded on board the German battleships. A faint white glimmer eastward delineated the sombre mass of the mountains, leaving the lowest slopes a band of shadow, reflected even more darkly in the still waters. In the darkness imposed by the mountains it was impossible to judge distances; it was only the stars that were fading.

The damp cool of early morning began to descend on the sea, spreading droplets of dew over the planks of Samuel's boat. I was dressed very lightly with nothing across my shoulders but an Albanian shirt made of thin muslin. I looked for my gold-embroidered jacket; it must be still in Aziyade's boat! A deadly chill spread up my arms into my chest. An hour to go until the best time to get back on board without attracting the attention of the watch. I tried to row but my arms would not be roused from their slumber. So, with infinite care I lifted the edge of the cloak that Samuel had over him and without wakening him I crept in beside this friend whom chance had bestowed on me.

And, in less than a second there were two sleepers, dead to the world, drifting along in a boat...

An hour later we were awakened by a raucous German voice shouting something like: 'Hey, you in that dinghy!' We had drifted right up to the German warships. With the guards' rifles trained on us we grabbed

the oars and made a hasty departure. Four o'clock; by the groping dawn

light I could make out the white mass that was Salonika and the dark

masses that were battleships. I crept back on board like a thief, only too glad to have escaped anyone's notice.

XXII

The following night (28th to 29th) I dreamt that, with no prior warning, I was having to leave Salonika and Aziyade. In the dream Samuel and I were trying to run along the road leading to the village where Aziyade lived so that I could at least say goodbye to her; then--as can happen in dreams--we were immobile; time slipped away and the corvette unfurled its sails.

'I'll send you some of her hair,' Samuel said, 'A whole long braid of her brown hair.'

And we kept trying to run. Then I was wakened for the next quarter watch.

It was midnight. The helmsman lit a candle in my cabin: I saw the light

catch the gilt woodwork and the silk flowers of the tapestry; then I was wide-awake.

That night it rained in torrents and I was soaked to the skin.

XXIII

Salonika, 29 July.

Out of the blue, at ten o'clock this morning, I received orders to prepare immediately to leave my ship and Salonika; to secure a passage on tomorrow's packet boat to Constantinople, and to join the British ship 'Deerhound', which was assigned to patrolling the waters thereabouts of both the Bosphorus and the Danube.

A party of sailors has just invaded my cabin. They are tearing down the hangings and packing my trunks.

My quarters on the 'Prince of Wales' were way below deck: a reinforced cubby-hole right next to the powder magazine. I had furnished this sunless vault in quite an original way: the iron walls were hung with a heavy red-silk tapestry displaying fantastic flowers; items of pottery, re-gilded bric-à-brac, and weapons glimmered from the darkness.

I had spent some sad hours in this gloom, hours you can't evade, the ones in which you come face to face with yourself and reflect with remorse upon the piercing regrets of the past.

XXIV

I had several good friends aboard the 'Prince of Wales' and they tended to indulge me, rather as they would a spoilt child; but I don't let myself become attached to people any more and parting from them doesn't affect me.

One more period of my life is coming to an end, and Salonika is one corner of the world I'll never see again.

And yet, I have known exhilarating nights on the tranquil waters of its great bay, nights for which many men would gladly pay a huge sum; and I

had come very close to loving this remarkably delightful young woman.

Soon I shall have forgotten those balmy nights when the first glimmers of dawn found us lying stretched out on a boat, drunk with love and bathed in morning dew.

I shall miss Samuel as well: poor Samuel, who so readily risked his life for me, and will cry like a baby when I depart. There I go again, still allowing myself to feel the tug of every strong affection, or what resembles such, be its motive selfishness or melancholy. Closing my eyes I'm willing to accept whatever feels like friendship or love if, for a single hour, it can fill life's terrifying emptiness.

XXV

30 July. Sunday.

Noon. The day is burning hot. I'm leaving Salonika. In the final hour Samuel rows up to the packet-boat to bid me farewell.

He seems remarkably casual and content,--another one who will quickly forget me.

'Au revoir, effendim, pensia poco de Samuel!' (Au revoir, monseigneur! Spare a little thought for Samuel!)

XXVI

'In autumn,' Aziyade told me, 'Abeddin-effendi, my master, is to move his household and his wives to Istanbul. If by chance he does not, I

will come by myself and find you.'

May it be so: I shall be waiting. But it will be a completely new beginning, a new way of life, in a new country, with new faces; and for how long, I do not know.

All the officers on board the 'Prince of Wales' made quite a show of waving me goodbye. Now the landscape, bathed in sunshine, is receding into the distance. For a long time it's possible to make out the white tower where Aziyade used to embark at night, also, dotted here and there with old plane trees, the stretches of stony ground I so often crossed in the dark. Salonika is soon no more than a grey smudge spreading over the arid mountain sides, a smudge bristling with tiny white marks, indicating minarets, and black ones, marking cypresses. And then the grey smudge vanishes, no doubt forever, behind the heights of Cape Kara-Bournou. The tops of four great mountains from mythology rise above the distant Macedonian coastline: Olympus, Athos, Pelion and Ossa!

BOOK TWO-SOLITUDE

I

Constantinople, 3 August 1876.

The crossing took three days; we called at Athos, Dedeagatch, and the

Dardanelles. My fellow passengers consisted of a beautiful Greek lady, two beautiful Jewish ladies, a German, an American missionary, his wife, and a Dervish. A somewhat odd assortment! But we got on very well with one another, and much of our time was taken up with music. General conversation was conducted in Latin or Homeric Greek. The missionary and I would exchange the odd comment in Polynesian. For the past three days I have been staying (at Her Britannic Majesty's expense) at a hotel in the Pera district. Next to me at table are a noble lord and his good lady. After dinner she and I spend the rest of the evening at the piano playing all Beethoven pieces. I'm in no hurry to see my new ship return from whatever part of the Sea of Marmora it's been patrolling.

II

Samuel, like the faithful friend he is, has followed me here. I was touched by that. He also had come by sea. He had managed to slip aboard one of the Messageries packet boats. He presented himself this morning and I embraced him heartily. I was so glad to see his open, honest face again—the only genial face in this great, wide city, where I don't know a single living soul.

'See, Effendi,' he said, 'I have left everything, my friends, my country, my boat, and have followed you.'

I'm already aware of the fact that it's among poorer people where one's more likely to find examples of total and spontaneous devotion; I much prefer them to people in easier circumstances because they have none of the latter's egoism or meanness.

III

All of Samuel's verbs end in '-ate'; everything that makes a noise is said to 'fate boum ('make boum').'

'If Samuel get on a horse,' he tells me, 'Samuel 'fate boum'.' (Read: 'Samuel will fall off.')

His thoughts come to him suddenly and in a jumble, as with a child. He is religious in a naïve and guileless way. His superstitions are eccentric and his observances preposterous. He's at his funniest when he adopts a serious demeanour.

IV

'My dearest brother,

'Like a bird evading capture, you fly off, you flit about, you land for a while then you're off again. Poor darling little bird, so capricious, so disillusioned, battered about by the winds, duped by every mirage, never finding a place where you might rest your weary head and your trembling wings. This time Salonika is the mirage-—but there will be others! Fly about in circles as you may, there'll come a time when these foolhardy urges will pall and you'll settle down for good on some pretty piece of greenery. No, you won't break your wings, and you won't fall into the abyss, because the God of little birds has spoken, and because there are angels watching over that dear thoughtless head of yours.

'It's no use my going on hoping: you won't be coming this year to sit beneath the lime trees. Winter will arrive and you still won't have set foot on our lawn. For five years I've watched our flowers come into bloom and shady places form; and I've harboured the thought that I would see you here, both of you. Through every season and every summer-time that was my dearest wish. Now there's only you, and you we shall not see.

'It's a fine August morning and I'm writing to you from Brightbury, from our rustic room that gives onto the courtyard, where the lime trees are; the birds are singing and everywhere the rays of sunlight filter everywhere through the leaves. It's Saturday and the ground and paving stones, now freshly watered, are whispering a country poem that would not, I know, leave you unmoved. The long spells of suffocating heat have gone and we are entering that period of peace and spirited charm that can so rightly be likened to a man's second age; the flowers and plants, all worn out by the sensual pleasures of summer, rise up again in vigorous reflowering, amid new verdure their colours show brighter, and a few fading leaves contribute to the virile charm of nature's second growth.

'In this little corner of my Eden, dearest brother, everything was waiting for you; everything seemed to be growing for you ... and once more it will all go by without you. It's simply a fact: we sha'nt see you.'

V

Taxim, a busy quarter on the heights of Pera: European carriages and clothes jostling with the carriages and costumes of the Orient: blazing heat and blazing sun: a mild breeze throws the dust and the yellowed leaves of August up in the air: the scent of the myrtles: the din of

the fruit sellers: streets cluttered with grapes and watermelons...The very first moments of my sojourn in Constantinople etch these images in my memory.

Being a total stranger, I would spend my afternoons beside the Taxim road, sitting in the breeze, under the trees. As I let myself drift back over the period that had

just ended, my eyes absently followed the cosi
stream in front of me. My thoughts were mostly
was surprised to discover that she was s
embedded in my mind.

It was in this part of the city that I became acquainted
with an Armenian priest who taught me the rudiments of
the Turkish language. At the time I had none of the
fondness for the country that I was to have

later: I gazed at it as a tourist would and I knew
almost nothing about Stamboul, a place feared by
Christians.

I remained in Pera for three months, thinking of ways
to carry out my impossible plans to live with her on the
other shore of the Golden Horn, to live the life of a
Moslem, to have her with me for days at a time, to
understand and penetrate her thoughts, to read what new,
wild and scarcely hinted-at notions lay deep in her heart,
and to have her all to myself.

My house in Pera was situated in a secluded spot
overlooking the Golden

Horn and the distant panorama of the Turkish city.
The splendour of summer lent charm to my abode; seated
by my large, open window, studying the language of
Islam, I would let my gaze hover above old Stamboul
lying bathed in sunlight. Away in the backgroud, in a
grove of cypress trees, Eyoub came into view; it would
have been heaven to be hidden with her there in that
mystical forgotten place where our life would have lit
upon its own strangely delightful setting.

All around my house, were immense stretches of land
with nothing but cypresses and tombs--empty terrain
where I spent more than one night with my mind bent on
careless adventures with Armenian or Greek girls. Deep

in my heart I was still faithful to Aziyade; but the days went by and she did not come ...

Of these lovely creatures all I retain is the charmless memory of passion-fed sensations; not a single thing links me with any of them, and they were quickly forgotten.

But I often wandered at night through these cemeteries and it brought me more than one regrettable encounter.

One morning, about three o'clock, a man came out from behind a cypress and barred my way. He was a nightwatchman; he was armed with a long iron-shod staff, a brace of pistols, and a dagger;--and I was unarmed.

I knew at once what this man wanted. He would sooner have killed me than abandon his plan.

I followed him obediently: I had a plan of my own. We were walking beside those gaping fifty-metre deep depressions which lie between Pera and Kassim-Pascha. He walked right along the edge; in an instant I seized an opportunity and flung myself at him;--out he stepped into the void, and over he went. I could hear him tumbling over rocks right down at the bottom; there was an ominous sound, then a groan.

He was sure to have companions nearby and in the silence of the night the sound of his fall would travel a fair distance. I fled into the night and I ran with so much speed no other human being could have caught me.

A white streak was appearing in the east as I returned to my room. The deathly pale face of debauchery often kept me down in the streets as late as this. I had no sooner dropped off to sleep when mellifluous sounds of music brought me awake; an ancient aubade, a song of lovers separating at dawn, a cheerful oriental melody, as fresh as daybreak, human voices accompanied by harps and guitars.

The choir went by and the voices faded away in the distance. All I could see through my wide-open window were the morning mist and the immense emptiness of the sky; then, high above, the pink shapes of minarets and a dome started to form; sketchily, little by little, the outline of the Turkish city appeared as if it were suspended in mid air... Then I recalled that I was in Stamboul,--and that she had sworn to come to me here.

VI

My encounter with the watchman had made a grim impression on me. I ceased my nocturnal wanderings and had no more mistresses,--except, that is, for a young Jewish girl who lived in Pri-Pacha, the Jewish quarter, and knew me by the name of Maketo. I spent the end of August and the beginning of September exploring the Bosphorus. The weather was mild and the sun shone brightly. The shady shores, the palaces and the mansions were reflected in the calm, blue waters criss-crossed with the wakes of gilded caiques. In Stamboul plans were in hand to depose Sultan Mourad and consecrate Abdul Hamid.

Constantinople, 30 August.

Midnight! the fifth hour, according to Turkish clocks; the nightwatchmen are striking the ground with their heavy, iron-shod staves. In the Galata quarter the dogs are in revolt and the howling down there is appalling. The dogs in this neighbourhood remain strictly neutral, and I'm obliged to them for that; they are asleep hodgepodge outside my door. All is peace and quiet. In the three hours I've spent stretched out by my open window I've been watching the lights go out one by one.

Below me the old Armenian cabin-like dwellings lie sleeping. From here I look out over a very deep ravine at the bottom of which, forming an impenetrably black shape, is a wood composed of centuries old cypresses; beneath these sad, balsam-scented trees lie ancient Moslem tombs. Together the minarets and the cupolas stand out against a star-strewn sky, wherein is suspended a slender crescent moon; the whole

horizon is fringed with towers and minarets--faintly outlined shapes, bluish against the paleness of the night. The big domes of the mosques, colourless as yet, climb, one on top of another, towards the moon, and in the mind assume gigantic dimensions. Down there, at this very moment, in one of the palaces, the Seraskierat, a dire comedy is being played out; the leading pashas have assembled there to depose Sultan Mourad; tomorrow Abdul Hamid will replace him. Sultan Mourad, whose succession three months ago prompted such festivity and who today is still treated like a god, will probably, tonight, in some corner of the Seraglio, be strangled.

Yet all is quiet in Constantinople... At eleven o'clock some cavalry and artillery went past my house at the gallop, heading for Stamboul; then from the batteries came muffled rumbling which petered out in the distance, and then everything fell silent again.

Owls are hooting in among the cypresses; they sound exactly as they do at home. I love this sound of summertime; it takes me back to woods in Yorkshire, to the beautiful evenings I spent under the trees at Brightbury.

Here, surrounded by all this stillness, images from the past come alive again, images of all that is shattered and gone, never to return.

I was counting on little Samuel being back this evening; I fear, though, that I'll never see him again. My heart is heavy and my loneliness is not easy to bear. A week ago, to enable him to earn a little money, I let him join the crew of a ship bound for Salonika. The three boats that might have returned him to me have come back without him. The last one arrived this evening and no one on board had heard anything of him...

The crescent moon is sinking slowly behind Stamboul, behind the domes of the Suleïmanieh. In this great city I am a stranger and unknown. Young Samuel was the only one who knew my name and that I existed. And I was beginning to be genuinely fond of him.

Has he too deserted me? Or has some mishap befallen him?

VIII

Having friends is like having dogs: it always ends badly. The best thing is to have neither.

IX

We have a friend, Saketo. He goes back and forth on the Turkish packet boats that run between Salonika and Constantinople. He visits us frequently. He was shy at first but he soon came to feel at home in my house. He's a sound fellow; he's been a friend of Samuel's since childhood. He brings Samuel news from home.

Old Esther, one of the Jewish ladies back in Salonika whose job it was to dress me up as a Turk and who called me her 'caro piccolo', sent me, Saketo reported, her kind wishes and remembrances. Saketi is always welcome here, especially when he brings me the messages that Aziyade

transmits to him through her negress.

'La hanum (the Turkish lady),' he says, 'sends her greetings to M. Loti. She asks him not to grow tired of waiting for her, and that before winter comes she will have arrived ...'

LOTI to WILLIAM BROWN

'I received your mournful letter only a couple of days ago; you addressed it to me aboard the 'Prince of Wales' so it pursued thence to Tunis and other places.

'So you too, old chap, are having to bear a heavy sorrow, and you feel it all the more keenly because your misfortune, like mine, is to have received the sort of education that develops one's heart and sensibility.

'You have doubtless kept all your promises regarding the young woman you're in love with. But for what good purpose? For whose benefit? And by virtue of what moral principle? If you love her to that extent and if she loves you too, don't worry your head about conventions and scruples; take her, whatever the cost, then for a while you will be happy; after that you'll be cured and everything else is of secondary importance.

'For the past five months, ever since I left you, I've been in Turkey. Here I've met an unusually charming young woman who has helped me while away my time during my exile in Salonika--and have met also a young vagrant called Samuel whom I regard as a friend.

'I spend as little of my time as possible on the 'Deerhound'; I come and go (like certain fevers in Guinea, reappearing every four days, to do what I'm supposed to do). I have a place of sorts to call my own in Constantinople, in a part of the city where no one knows me; there I lead a life that is ruled only by my whim, my current mistress being a little Bulgarian girl of seventeen summers.

'The Orient still has its charm; it has stayed more oriental than you might imagine. I have performed the mighty feat of learning Turkish in

two months. I've adopted the fez and the kaftan,--and in the same way children play at being soldiers, I play at being an effendi, a gentleman. Time was when I mocked novels that told of 'worthy people',

who, following some disaster or other, lost their sensibilty and moral understanding,--though perhaps my own case is not so different from theirs. Now I'm done with suffering and being tied to memory: I could pass by on the other side those whom I once adored.

'I tried being a Christian, but it was no good. This sublime illusion, which is able to heighten the courage of some men and some women--our mothers for example-- enough to make them heroic, this illusion is denied me. The Christians of this world make me laugh; if I were one my view would be that nothing but that counted any more; I would turn missionary and go off somewhere and get myself killed following Christ...

'Take it from me, old chap, time and debauchery are two marvellous remedies. In the end the heart becomes numb and it's then when one ceases to suffer. This truth isn't new and I recognise that Alfred de Musset could have phrased it better for you; but of all the old sayings that men hang on to generation after generation, it is one that will never die out. This pure love you dream of is a fiction, like friendship; forget her you love and settle for some shameless man-hunter. If your ideal woman eludes you, go and fall for a circus girl, one with a beautiful figure.

'There is no God, there are no morals, and of all that we were taught to respect nothing remains; our lives don't last very long, hence it's reasonable to seek all the pleasure we can while we wait for the final terror which is death. The truly bad things are sickness, ugliness, and old age; neither you nor I are afflicted by them; we can still have lots of mistresses and enjoy life. I am going to open my heart to you: my rule of life--despite all moral principles and all social conventions--is to always do what pleases me. I believe in nothing and nobody; I love nobody and nothing; I possess neither faith nor hope. It has taken me twenty seven years to reach this point; if I have fallen lower than the average person it's because I started from a greater height.

'Accept my very fond farewell.

'LOTI.'

XI

The mosque of Eyoub is situated at the far end of the Golden Horn; it was built at the time of Mahomet II over the tomb of Eyoub, companion of the Prophet. Access to it has always been forbidden to Christians and even its immediate surroundings are none too safe for them.

The building is of white marble; it stands in a lonely place out in the country and is surrounded by cemeteries. Its dome and minarets are hard to see above the dense greenery of a clump of gigantic plane trees and ancient cypresses.

The paths in the cemetery, mainly sunken, are paved with stone or marble, and lead through deep shade. On either side stand age-old marble buildings whose pristine whiteness contrasts with the black-tinged cypresses.

Surrounded by flowers hundreds of gilded tombs are clustered about these shaded walks; these are the tombs of the venerable, of pashas long gone, and of renowned Moslem dignitaries. Down one of these mournful avenues can be found the tomb-pavilions of the learned Sheiks of Islam.

It's in the mosque at Eyoub where the Sultans are consecrated.

XII

On the sixth of September at six o'clock in the morning, I managed to get myself into the second inner courtyard of the mosque at Eyoub.

The ancient memorial was silent and empty. The two Dervishes I had with

me were trembling from head to foot at the audacity of such an enterprise. Without a word we walked across the marble flagstones. At that early hour the mosque was as white as snow; in the deserted courtyards hundreds of wood pigeons carried on with their pecking and fluttering about. The two Dervishes, robed like monks, lifted the leather hatch closing off the sanctuary and gave me a clear view of this venerated place, the holiest in Stamboul, one that no Christian eyes had ever beheld.

That was on the day before -the consecration of Sultan Abd-ul-Hamid.

I remember the day when the new Sultan, amid great pomp, went to take possession of the imperial palace. I had been one of the first to see him as he left the old Seraglio, that gloomy retreat designated the residence of heirs to the throne. Great ceremonial caiques had come to convey him, and, for a brief moment, my own caique touched his.

Those few days of power had already aged the Sultan. He no longer wore that look of youth and energy. The extreme simplicity of his attire contrasted with the over-abundant oriental trappings of his new environment. This man who had been brought out of comparative obscurity to assume supreme power appeared to be deep in a troubled reverie; he was thin and pale and had a vague, melancholy air about him; there were dark brown rings round his deep black eyes; his features betrayed intelligence and fine breeding.

Carved and gilded all over and bearing a golden spur at the prow the Sultan's caiques, each with its twenty-six oarsmen, retained the elegant contours of their oriental forerunners. I remember the court livery: green and orange covered with gold trimmings; and the Sultan's throne decorated with several suns and standing beneath a red and gold canopy.

XIII

Today, the seventh of September, I witnessed the grand spectacle of the Sultan's consecration.

Abd-ul-Hamid, so it seems, is in some hurry to surround himself with the glamour of the Khalifs; it could be that his accession will open up a new era in Islam and bring Turkey a little more glory and a final burst of splendour. He entered the holy mosque of Eyoub where, amid great pomp, he girded on the sword of Osman; afterwards, with a long and magnificent procession following him, the Sultan, made his way to the palace of the old Seraglio, traversing the whole length of Stamboul, pausing, as is customary, to offer a prayer in the mosques and funeral pavilions he came upon.

At the head of the march were the Halbardiers, clad in scarlet that was richly ornamented in gold, and wearing green plumes two metres high. Abd-ul-Hamid mounted on a stupendous white horse with gold and gems on its trappings rode in their midst, maintaining a pace that was slow and majestic. Then came the Sheikh-ul-Islam in a green mantle, the Emirs in cashmere turbans, the Suleimans in white turbans with strips of gold, and the grand dignitaries upon horses that glittered with gold--a solemn,and interminable procession that revealed any number of strange-looking faces! I remember that the eighty year old Suleimans, supported by lackeys and riding quiet mounts, were distinguishable by their white beards and by their sombre looks that were fraught with impenetrable fanaticism.

Surging the whole way along the route was a huge crowd, one of those

Turkish crowds--compared with which the most extravagant western crowds

appear ugly and dowdy. The platforms that had been erected along a stretch of several kilometers were bending under the weight of those who were keen to see everything; and all the varied European and Asian styles of dress intermingled with one another. On the higher parts of Eyoub moved a throng of Turkish ladies, each individual shape enveloped down to the feet in a length of brightly coloured silk; the

white head coverings were hidden behind the folds of the yashmaks from which dark eyes looked out. From a distance the figures beneath the trees could be mistaken for the painted and storied gravestones. So colourful and bizarre was the scene it might be better described as the fantastical imagining of some hallucinating orientalist.

XIV

Samuel's return has brought a little cheer to my dreary abode. Fortune smiles on me when I'm at the roulette wheels in Pera, and autumn in the Orient is magnificent. I'm living in one of the loveliest countries in the world and I'm free to go wherever I choose. I'm able to explore villages, mountains and woodland along the coasts of both Europe and Asia and it would take many a poor soul a whole year to see what I see and have the adventures I have in a single day.

May Allah grant long life to Sultan Abd-ul-Hamid who has revived the great religious festivals and ceremonies of Islam. Every night Stamboul is illuminated and Bengal lights are strung along the Bosphorus--the last glow of an Orient that is disappearing, never to return.

Despite my indifference to politics, my sympathies reside with this beautiful country, which some foreign powers wish to abolish. Very gradually and without my knowing it I am becoming a Turk.

XV

Concerning Samuel and his nationality: he is Turkish by chance, Jewish by religion, and Spanish by parentage.

In Salonika he was part vagrant, part boatman-porter. Here, as there, he picks up jobs on the quays. As he's better looking than the others he gets plenty of work, and most days does well; his evening meal is a bunch of grapes and a hunk of bread; and then he returns to the house, happy to be alive.

My luck at roulette has run out and we are extremely poor, both of us; but being so carefree makes up for it; and besides, we are young enough to be able to get for free what others have to pay a high price for.

Before he sets off for work Samuel puts on two pairs of trousers that have holes in them; by his reckoning the position of the holes won't coincide, with the result that public decency will not be offended.

Evening finds us, like a couple of proper orientals, outside a Turkish café, under the plane trees, smoking our hookahs; or we may go along to the Chinese shadow-theatre to see Karagueuz, the Turkish equivalent of Punch, who has us spellbound. We keep clear of all the unrest that's about: as far as were concerned, politics don't exist.

There's panic, however, among the Christians in Constantinople and Pera folk live in terror of Stamboul and no longer cross the bridges to it without trembling.

XVI

Yesterday evening I went on horseback over to Stamboul. I wanted to visit Izeddin-Ali. It was during the important festival of Bairam, an

oriental extravaganza and the last commemorative holiday of Ramazan: all the mosques were illuminated, each minaret sparkling right up to its topmost point; suspended in the air were verses from the Koran done in luminous letters; at the sound of the canon thousands of men shouted with one voice the holy name of Allah; in the streets crowds of gaily dressed people paraded with torches and lanterns; groups of women moved around, dressed in silk adorned with silver and gold.

At three o'clock in the morning, having covered the whole of Stamboul,

Izeddin-Ali and I fetched up at an underground den on the outskirts of the city. Young Asian boys, dressed up as belly-dancers, performed lascivious dances before an audience composed of every ex-convict from the

Ottoman jails. It was an orgy whose novelty sickened me. I begged off seeing the last part which would not have been out of place in Sodom in its heyday. We got back to my friend's house early in the morning.

XVII

Karagueuz

The adventures and misdemeanours of His Lordship Karagueuz have entertained countless generations of Turks, and there is nothing to indicate that the popularity of this individual is nearing its end.

Karagueuz shares many of the traits of the old French character Polichinelle; having given everyone, including his wife, a good hiding, he in turn receives a beating from Satan, who finally carts him off, much to the delight of the spectators.

Karagueuz is made of cardboard or wood; he appears in the form of a puppet or a moving silhouette; he's equally funny as either. He adopts voices and postures Punch has never dreamed of; the caresses he lavishes upon Madame Karagueuz are irresistibly comic. Sometimes he will bawl out to the spectators and set himself at odds with them. Sometimes he's facetious (albeit in quite an ill-chosen way), and in plain view of every one he does things that would scandalise even a monkey. In Turkey it's all quite acceptable and in order, and each evening one can see decent family folk, lantern in hand, taking troupes of young children to see Karagueuz.

So it is,then, that roomfuls of infants get to watch a show that would prompt blushes in an English guardhouse. It's just one of those oddities in oriental behaviour: it would be completely erroneous to be even tempted to infer that Moslems are far more depraved than we are.

Karagueuz theatres open on the first day of the lunar month of Ramazan and for thirty days they are all the rage. At the end of the month everything is dismantled and put away. For a whole year Karagueuz stays in his box and is not let out out, not any pretext whatsoever.

XVIII

Pera bores me, so I'm moving away. I'm going to live in Old Stamboul-- that's to say, the far side of Stamboul in the holy suburb of Eyoub.

Over there, where no one knows my real name or my social standing, I go by the name of Arif-Effendi. My kind Moslem neighbours are under no illusion as to my nationality; but that doesn't bother them, nor me.

It takes me about two hours to get to the 'Deerhound'. I'm almost out

in the country and I have a house all to myself. This part of Eyoub is Turkish and is picturesque as can be: there's a village street that bustles with life all day long; there are bazaars, cafes, tents, and serious-looking dervishes who sit beneath the almond trees and smoke their hookahs.

There is a square, graced by an old monumental fountain, made of white marble, which is the meeting point for all those who arrive here from

the interior: gypsies, travelling acrobats, bear-trainers. On the square stands a solitary house--this is ours.

On the ground floor: a hallway, whitewashed with lime and as white as snow, and a set of rooms that are empty. (We never open that part of the house, except in the evening, before going to bed, to check that there's no one hiding there; Samuel thinks down there is haunted.)

On the first floor: my bedroom with its three windows overlooking the square that I mentioned earlier; then Samuel's room and the women's room with its view eastwards towards the Golden Horn.

Up another flight of stairs and one is on the roof which, like other Arab roofs has a terrace shaded by a vine whose leaves, alas, have already been turned completely yellow by the November wind. Standing right next to the house is an old village mosque. When my friend the muezzin* climbs to the top of his minaret he's on a level with my terrace and before he begins to chant his prayer he always greets me with a friendly salaam.

The view from up there is beautiful: the bleak stretch of country lying between Eyoub and the farthest point of the Horn, the sacred mosque with its marble whiteness emerging from the mysterious depths of a grove of ancient trees, then some mournful hills touched with sombre colours and strewn with the marble of vast cemeteries, truly a city of the dead.

To the right the Golden Horn, criss-crossed by thousands of gilded caiques; then the whole of Stamboul, foreshortened, its mosques all tangled up, their domes and minarets a blur.

Much farther out, a hill covered with white houses. This is Petra, the Christian town, and behind it the 'Deerhound' lies at anchor.

XIX

When I first set eyes on the empty house I felt very discouraged: bare walls, windows coming apart, doors with no locks; and besides, it was so far from the 'Deerhound'. The whole idea was impractical.

XX

Samuel has spent eight days cleaning, whitewashing, and stopping up gaps. For carpeting we're having white mats nailed to the whole floor, a Turkish practice that's clean and comfortable. We've got curtains hanging at the windows, and a wide divan, covered in a red leafy pattern. In sum the first, albeit for now modest, part of our moving in. Already the house looks different; I can foresee the possibility of making this my home--draughts and all; I'm finding the idea less daunting. However, it would need the presence of the one who has sworn to come, and perhaps it is for her sake alone that I have cut myself off from the world.

Here in Eyoub I tend to be indulged rather like the local spoilt child. Samuel too is quite a favourite.

My neighbours, wary of me in the beginning, eventually decided to shower kindnesses upon the charming foreigner whom Allah had sent and whose domestic arrangements were deeply puzzling.

At the end of a two hour visit the Dervish, Hassan-Effendi, summarised his conclusions: 'You are a most extraordinary fellow and everything you do is odd. You are very young, or at least you appear so, and you live a life of complete independence, the like of which a man of mature age cannot always command. We do not know where you come from and you have no visible source of income. You are already familiar with every part of the globe. You have a mass of knowledge surpassing that of our Ulema; you know everything; you have seen everything. You are twenty years of age, perhaps twenty-two, yet one human lifetime would scarcely be enough for all your mysterious past. Your place would be in the first rank of European society in Pera, yet you come and live in Eyoub with a Jewish vagabond as your choice companion. You are a most extraordinary fellow but it is a pleasure for me to see you here, and I am delighted that you have come to settle among us.'

XXI

The Ceremony of Surre-i humayoun; Imperial gifts go off to Mecca

The Sultan each year sends to the Holy City a caravan loaded with offerings.

The procession sets off from the palace of Dolma-Bagtche and makes its way to the quay at Top-Hane, there to embark for Scutari on the Asian shore.

At the head of the procession is a group of Arabs dancing to the sound of the tomtom and shaking aloft long poles wound round with golden streamers.

Camels with ostrich feathers on their heads advance with solemn tread, carrying on their backs a framework covered in gold brocade and gemstones and containing the most precious of the gifts. Plumed mules carry the rest of the Khalif's tribute in boxes covered with red velvet, embroidered with gold.

Coucillors and leading dignitaries follow on horseback. Troops form a guard of honour along the entire route. It's a forty day march from Stamboul to the Holy City.

These November nights descend like a pall over Eyoub. During those first nights that I spent all on my own in the house my heart was heavy and I was beset by any number of worrisome feelings.

When I first closed my door against the invading darkness a deep sadness wrapped itself around me like a shroud.

I thought I'd venture outside. I lit a lantern. (It means jail to be caught wandering about Stamboul without a lantern.)

In Eyoub, after seven o'clock in the evening, everywhere is locked up; when the sun sets the Turks bolt their doors and go to bed.

Every now and again a lamp may throw the shadow of a barred window onto the pavement. It's best not to peer through the window. The lamp will

be a funerary lamp and all it illuminates will be great catafalques

surmounted by turbans. A fellow could be in danger of getting his throat cut on the pavement but there's no one to come out and help. The lamps, flickering in there till dawn, are less reassuring than the darkness.

In Stamboul these corpse dwellings are at every street corner.

And quite close to our own house, where the streets end, is

where the great cemeteries begin. They are the haunt of gangs of criminals who, when they've robbed you of everything you have, will bury your dead body right there, without the police ever getting themselves involved.

A night-watchman wanted to know why I was taking a stroll at that hour. My answer seemed to mystify rather than satisfy him. He made me promise to return to my house.

Fortunately there are some really decent fellows among the nightwatchmen, and this one in particular, who would see all sorts of mysterious comings and goings, remained a model of discretion.

XIII

'One can find a companion, but not a faithful friend.'

'One can travel the whole world over and yet never find a friend.'

Extracts from an old Oriental poem.

LOTI to his SISTER in Brightbury

Eyoub..., 1876.

'Opening my heart to you is becoming more and more difficult because every day your point of view and mine move further apart. The Christian notion lingered in my imagination even when I no longer believed; it had a vague, consoling charm. Today its allure is gone completely. I know of nothing that is so futile, so misleading, so unjustifiable.

'There have been some dreadful moments in my life, and I have suffered cruelly, as you know.

'I told you once that I wanted to get married and I entrusted to you the task of finding a young woman who would suit our ancestral home and our our ageing mother. I am asking you now to give no more thought to the matter. I would make the woman I married unhappy; I would rather continue my life of pleasure.

'I'm writing to you from my dreary house in Eyoub. Apart from a young lad by the name of Yousouf, who has had to get used to my giving orders to by sign (so loath am I to even speak), I spend long hours here without addressing a single living soul.

'I've told you that I have no confidence in another's affection; it's true. I have friends who show me a great deal of affection, but it's not something I trust. Samuel, the friend who has just left me, is still, perhaps, the one out of them all who cares for me most. But I'm under no illusion: his affection is the enthusiastic admiration of a child. One of these fine days it will all vanish, just like smoke, and I'll find myself alone again.

'As for your own affection, dear sister, I believe in it to a certain extent; at its very least it has the force of habit; and then again one has to believe in something. If you really love me, tell me, show me ..... I need to be bound to someone once more; and if it's true, help me to believe it. I feel the ground crumbling away beneath my feet, everything around me is becoming meaningless and I'm deeply distressed ...

'In so far as it will help keep my dear old mother alive I will remain outwardly the same as I am today. When she is no longer with us I will come and bid you farewell; then I will disappear and leave not a trace behind ...'

XXV

LOTI to PLUMKETT

Eyoub, 15 November 1876.

'Behind all this oriental phantasmagoria surrounding my existence, behind Arif-Effendi, there is a poor, weary fellow who often feels a mortal chill at his heart. There are few people this fellow (very reserved by nature) can occasionally have a close friendly chat with, but you are one of them. Do what I may, Plunkett, I'm not happy, and I can't find any way to rid myself of my unhappiness. My heart is all weariness and bitterness.

'Being so alone I have become attached to a barefot bundle of rags off the Salonika quays. He's what the late Raoul de Nangis would have called 'a rough diamond in an iron setting'; he's without guile and is eccentric; I am less bored when he's around.

'Outside, as I write, the depressing winter twilight is upon us. All that can be heard in the neighbourhood is the muezzin sadly chanting, in honour of Allah, an age-old lament. Images from the past rise up in my mind with painful clarity. Objects around me take on a look that is sinister and disturbing, and I wonder what I'm doing here in my

village retreat in the middle of nowhere.

'If only she were here,--she: Aziyade.

'My waiting never ceases,--alas, neither did Sister Anne's.

'Now that I've closed my curtains, lit my lamp and my fire, the setting is altered; likewise my thoughts. I continue writing my letter, sitting before a merry blaze, wrapped in a fur coat, with my feet on a thick Turkish rug. For a moment I could have sworn I'd turned Dervish--an entertaining thought.

'I don't really know what I can tell you about my life that would amuse you; there's plenty to tell; but choosing what, is the difficulty. And, then, what's past is past, is it not?--and no longer interests you.

'Several mistresses (none of whom I actually cared for), lots of incidents, lots of wandering on foot or on horseback across mountains and valleys; everywhere faces that were unfamiliar, unconcerned or

disagreeable; lots of debts; Jews in hot pursuit; gold-embroidered outfits that skim the ground; death in my soul and and an emptiness in my heart.

'That is how things are with me this evening, fifteenth of November, at ten o'clock. It's winter; a freezing shower and a strong wind beat against the window panes of my drab house; that's the only sound that can be heard; and at this hour the old Turkish lamp hanging above my head is the only left burning in Eyoub. Eyoub, the heart of Islam, is a depressing area; in it stands the sacred mosque where the Sultans are consecrated. Some fierce old Dervishes and the guardians of the holy tombs are the only inhabitants of this quarter, which is the one that is most deeply Moslem and fanatical...

'As I told you, your friend Loti is in his house all by himself, well wrapped up in a coat of fox fur, and assuming the role of a Dervish.

'He has bolted the doors and is enjoying the selfish comfort of being snug at home, all the more so with this gale outside and the area itself being unsafe and inhospitable.

'Like all things that are amazingly old, Loti's room affords one strange dreams and deep contemplations. Down the years its carved oak ceiling must have given shelter to some curious occupants and viewed many a drama.

'I've kept to authentic colours throughout. The floor is covered over with woven mats and Turkish carpeting, the one luxury in the house; and, following the Turkish custom, shoes are removed as one enters, to avoid bringing in any dirt.

'Practically all there is by way of furniture consists of a very low divan and cushions lying about the floor, marks of the sensual informality of the people of the East.

Weapons and ages-old decorations hang from the walls which carry painted verses from the Koran mingled with images of flowers and extraordinary animals.

'Adjoining my room is the part of the house reserved for the women, the 'haremlike', as we say in Turkish. Empty now; it too is waiting for Aziyadé, who would be with me already, if she'd kept her promise.

'Another little room next to mine is likewise empty: it's Samuel's room; he's gone to Salonika for me to get news of the girl with the green eyes. But it seems he's no more likely to come back to me than she is.

'If she were not to come, however,then, by God, it wouldn't be long before another took her place. But it wouldn't be anything like the same. I was almost in love with her, and it's for her that I've become a Turk.'

XXVI

To LOTI from his SISTER

Brightbury ..., 1876.

'My dearest Brother,

'Since yesterday when I read your letter I have been unable to shake off any of the despair it caused me ... You wish to disappear!... One day, soon perhaps, when our much loved mother leaves us, you want to disappear, to abandon me for ever, to make 'tabula rasa' of all our memories, to bury our past,--the old family home at Brightbury sold, things that we treasure dispersed,-- and there you will be--not dead...! you'll be someplace in the clutch of Satan, stagnating, some place I don't know, some place where I shall feel you getting old and in pain!... Better for God to bring about your death. Then I shall grieve for you; then I shall know that the emptiness is meant to be; I shall accept that, and I shall bow my head.

'What you are saying appalls me and tears at me. Yes, you could do it: your very words tell me that. And you could do it without any sign on your face or twinge in your withered heart because the path you've set yourself on is wicked and damned, because I no longer count for anything in your existence....Your life is my life, there is a corner hidden in my heart that's for no one else but you, and when you go from me it will be empty and will consume me.

'I have lost my brother, I am told in advance—only a matter of time, a few months, perhaps,--he is lost for all time and eternity, he has

already died a thousand deaths. Everything is collapsing, everything is breaking in pieces. Behold the precious child plunging into a bottomless pit,--the bottomless pit! He is in pain, he has no air, no light, no sun; but he is powerless; his eyes remain fixed at the depths and at his feet; he no longer looks upward, he no longer can, the Prince of Darkness forbids him to...Yet there are times when he wants to. He can hear a distant voice; it's the voice that lulled him to sleep as a child; but the Prince tells him: 'It's lying, it's insane, futile!' and the poor child, bound, half-throttled, in the depths of its pit, bleeding, frantic, having been taught by his master to call good evil and evil good, what does he do?...he smiles.

'Nothing surprises me that comes from that poor overwrought soul of yours, not even Satan's mocking smile... He has no option!

'You have even lost, poor brother, that hunger for straightforward living you spoke to me about. You no longer have any desire for that gentle and unassuming, youthful, tender and pretty, loving companion, mother of the children you would have adored. I used to picture her there, in the old drawing room, seated beneath the old portraits....

'A wind of corruption has passed over all that. This brother whose heart cannot even live without affections, who hungers and thirsts after them, will have no more to do with pure affections; he will grow old, but there will be no one there to cherish him and win his smile. His mistresses will laugh him away; there's nothing more they can give; and then, abandoned, all hope gone.... he will die!

'The more unhappy you are, the more miserable, undecided, assured you are, the more I love you. O my beloved darling brother, if only you wanted to have a life again! If only God would will it! If only you could see the desolation of my heart, if only you could feel the warmth of my prayers.

'But there's the fear, the tedium of conversion, the pallid terrors of the Christian life ... 'Conversion', what a horrid word!...Boring sermons, absurd people, dull Methodism, austerity without light and colour. Long words, and Canaan patois!...Will all that entice you? None of that, in fact, is Jesus; and Jesus to you is not the radiant Master whom I know and worship. From Him you will receive neither fear, nor boredom, nor turning away. In your strange sufferings, your consuming sorrows, He will weep with you.

'Darling brother, every hour of the day I pray for you; never has my heart been so full of thoughts of you... It may not be in ten years, in twenty years, but one day, I know, you will come to believe. Perhaps I'll never know,--perhaps it won't be long before I die,--but I will go on hoping and praying.

'I think I've written far too much. So many pages! not easy to read! My darling is already raising his shoulders in a shrug. Will there come a day when he no longer reads anything from me?'

XXVII

'Old Kairoullah,' I said, 'bring me some women!'

Old Kairoullah was sitting on the ground in front of me. He was curled up like a nasty, vile insect; his bald, pointed head gleamed in the faint light from my lamp.

It was eight o'clock on a winter's evening and all Eyoub was as black and silent as a tomb.

Old Kairoullah was raising a twelve year old son, beautiful as an

angel, called Joseph, whom he adored. In all other respects he was an unrivalled scoundrel. A Jew with no standing, he had a hand in all the seedy dealings in Stamboul; one in particular brought him into contact

with Yuzbachi Suleimanas and several of my Moslem friends.

However, he was accepted and tolerated everywhere by reason of the fact that as the years passed people had grown used to seeing him around. When one met him in the street one would say: 'Good day, Kairoullah!' and even touch the tips of his big, hairy fingers.

Old Kairoullah mulled over my request and replied:

'Monsieur Marketo, right now women are very expensive. But,' he added, 'there are less costly entertainments that I can offer you this very evening, Monsieur Marketo... A little music, say, will doubtless suit you...'

With this oddly worded proposal he lit his lantern, put on his fur-trimmed coat, his footwraps, and disappeared.

Half an hour later the door curtain in my room rose to admit six young Jewish boys wearing fur-trimmed robes, red, blue, green, and orange. Kairoullah escorted them, together with another old man, more hideous than he. After some exaggerated bowing and curtseying the whole bunch sat down on the floor. I remained as impassive and motionless as an Egyptian idol.

The children carried little gilt harps which they began strumming with fingers laden with tawdry rings. What resulted was a music like no other; I listened for a few minutes in silence.

Old Kairoullah bent down and said in my ear: 'One can see they are much to your liking, Monsieur Marketo.'

I had read the situation already and I betrayed no surprise; I was simply curious to take my study of human abasement a little further.

'Old Kairoullah,' I said, 'your son is better looking than they are ...'

Old Kairoullah paused for a moment and then replied:

'Monsieur Maketo, we can talk about that again tomorrow...'

... When I'd shoo'd them all away like you would a pack of mangy looking animals, I saw the door curtain lift and old Kairoullah's head re-emerge soundlessly and crane forward.

'Monsieur Maketo,' he said, 'have pity on me! I live a long way from here. I'm taken for a rich man. Better to kill me with your own hand than to turn me out at such a late hour. Let me sleep in some corner of your house, and I'll be on my way before daybreak, I swear.'

I couldn't have done that to the old man; even if he'd not been murdered, he'd have died from cold and fear. So I indicated one of the corners, and there he remained throughout the whole freezing night, in his threadbare coat and curled up like an old woodlouse. I would hear him shudder; he would let out coughs from deep in his chest and they sounded like moans. I had enough pity to get up again and throw him a rug to use as a blanket.

At first light I ordered him to clear off, and I advised him never to set foot in my house again, nor to even think of approaching me for any reason whatsoever.

BOOK THREE-TWO IN EYOUB

I

Eyoub, 4th December 1876.

I was told: 'She's arrived!'--and for the next two days I lived in a fever of expectation.

Then I was informed by Kadidja, the old negress who, back in Salonika used to accompany Aziyade in her boat at night and risk her life for her mistress, that that evening a caique was to take Aziyade to the jetty at Eyoub, which was opposite my house.

I waited for her there for three hours.

It had been a bright, sunny day; traffic on the Golden Horn had been uncommonly busy; as the day drew to a close thousands of caiques were arriving at Eyoub jetty, bringing back to the peace and calm of their own neighbourhood Turks who worked in the crowded centres of Constantinople, Galata and the Grand Bazaar.

I was beginning to be known in Eyoub. I'd hear:

'Good evening, Arif; and what are you waiting for?'

They all knew very well that my name couldn't be Arif, and that I was a Christian from the West; but no longer did anyone cast long looks on my Eastern whimsy, and my chosen name stayed.

II

'Portia! Bright flame from heaven! Portia!

give me your hand, it is I!'

(ALFRED DE MUSSET, 'Portia.')

Two hours after the sun had set, a final, solitary caique, coming from Azar-Kapou, drew close; Samuel was at the oars; seated on cushions at the stern I could see the veiled form of a woman. It was her.

When they arrived the square round the mosque had become deserted and the night had grown cold.

I took hold of her hand and ran with her to my house, forgetting all about Samuel who was left outside...

And, when my dream became reality, when she was there in the room I had got ready for her, alone with me, behind two iron-studded doors, all I could think of doing was dropping to the floor beside her and hugging her knees. I felt I had longed for her so deeply and for so long there was nothing left of me.

Then I heard her voice. That was the first time-- rapturous moment!--she had ever spoken and I had been able to understand her. Yet I was bereft of all the Turkish I had learnt purposely for her; I answered her in a garbled old-fashioned English that even I couldn't follow.

'Severim, seni, Loti!' she said. ('I love you, Loti, I love you!')

I had had these words, these timeless words, said to me before Aziyade; but this was the first time love's sweet music, rendered in the Turkish tongue, had struck my ears. Exquisite music that I'd forgotten. Possibly I still hear it rising with such intoxicated abandon from the depths of a young woman's pure heart. I don't think I've ever heard anything like it, ever heard any song that could so fill my empty soul.

Then I gathered her up into my arms, and turned her so I could see her face in the lamplight.

'O speak again!' I said, like Romeo. 'Say those words once more!'

And I started saying all manner of things to her, knowing she would understand me; I was making sense at last, and I was using Turkish; I asked her a host of questions, and kept insisting:

'Say something.'

But she simply gazed at me in ecstasy and I saw that she couldn't fathom what I was saying, and that my efforts were in vain.

'Aziyade,' I cried, 'don't you understand me?

'No,' she answered.

And in a low voice she said with a directness that was both tender and wild:

'I wish I could eat the words that come from your mouth! Senin laf yemek isterim! Loti! I wish I could eat the sound of your voice.'

III

Eyoub, December 1876.

Aziyade is not one to talk much; she'll often smile, but never laugh; her footsteps betray not a sound; her movements are supple, undulating, leisurely, silent. Yes, that's the weird little creature who more often than not will vanish when dawn breaks and reappear at nightfall when jinees and ghosts are around.

There's something of an aura about her too, and seems to bring light wherever she goes. You expect to see rays of light encircling her serious, child-like head; and you do indeed see them when the light catches some of those tiny, delicate strands that defeat attention and exquisitely enclose her cheeks and brow.

She considers these tiny hairs as being regrettable and spends an hour every morning in entirely unsuccessful efforts to smooth them down. This task and the other one, which consists of colouring her nails an orange red, are her two principal occupations.

She's idle, like all women brought up in Turkey; however, she knows how to embroider, make rose water, and write her name. She writes all over my walls with such earnestness you'd think it were a matter of some importance; and she blunts all my pencils doing it.

Aziyade communicates her thoughts to me more with her eyes than her lips; it's amazing how greatly their expression varies and and how agile they are. She is so accomplished in this silent language that she could speak a whole lot less than she actually does, or even dispense with it all together.

In certain situations it seems her best recourse is to reply by singing snatches from one or two Turkish songs. Such a method would, in a European woman, denote a lack of taste; but, in Aziyade's case it takes on an unusual Turkish charm.

Though very young and fresh her voice is deep; low-pitched notes always suit it, and at times the Turkish aspirates render it slightly husky.

Aziyade is eighteen or nineteen years old. She is capable, suddenly and of her own accord, of making drastic resolves, and, cost what it may,

of seeing them through to the end.

IV

Back in our Salonika days, when I risked Samuel's life and my own for the sake of being with her for a single hour, I'd been nursing this crazy dream: living with her in some hidden corner of the East, somewhere where poor little Samuel could find us. Contrary to all Moslem teaching and impossible in every respect, I have virtually made my dream a reality.

Constantinople was the only place where anything like that could have been attempted; it's a veritable human desert of which Paris was once the classic example: a collection of several large towns where everyone could live as they pleased, free from restrictions, and could adopt several personalities, all different--Loti, Arif, and Marketo.

Let the winter wind blow; let the gusts of December rattle the ironwork on our doors and the bars at our windows. Protected by heavy iron bolts, by a whole arsenal of loaded weapons,--by the inviolability of Turkish dwellings,--and sitting before a copper brazier... little Aziyade. How good it feels in this home of ours!

V

LOTI to his SISTER at Brightbury.

'My dear little sister,

'I've been obstinate and ungrateful in not writing to you earlier. I've hurt you grievously, you say, and I can believe it. Unfortunately I meant everything I wrote, and I still do; there's nothing I can do now to remedy the pain I brought you; I simply made the mistake of revealing to you all that lies deep in my heart, but that is what you wanted.

'I believe you love me; if I needed proof your letters alone would assure me. You know I love you too.

'You think, then, that I should cultivate an interest in something, something fine and honourable, and take it seriously? But I do have my dear old mother; she is now an object in my life, the one object of my devotion. For her I attempt a degree of cheerfulness and courage: for her I preserve the positive and sensible side of my existence—I become Loti, naval officer, again.

'I agree with you, I can think of nothing more hideous than an old libertine, worn, weary, and left to die all alone. But that is what I shall not be: when I am no longer healthy, young, and loved, that is when I'll disappear.

'I think you've misunderstood me: I shall disappear because I shall be dead.

'When I come home again I will make one final effort for both of you, but for you especially. Back with my family my ideas can change; if you

decide upon a young woman that you're fond of, I will do my best to love her, and out of my love for you, to content myself in that attachment.

'As I've written to you before about Aziyade, I'm now happy to be able to tell you that she has arrived. She loves me with all her soul, and doesn't think it possible that I would ever leave her. Samuel is back

as well. The two of them surround me with so much love that I forget the past and the bad times,--and to some extent the times that have been missed...'

VI

From its modest beginnings, Arif-Effendi's house has, bit by bit, became a house of luxury: carpets from Persia, door curtains from Smyrna, pottery, weapons--all these fine things have arrived one by one, not without difficulty, but this mode of collecting has added to their charm.

Roulette has provided us with blue satin hangings that

are embroidered with red roses and come from some seraglio; and the walls that were formerly bare, are now decorated with silk. Luxury such as this, concealed in an isolated hovel, seems to come from the realm of fantasy.

Aziyade too brings home some item or other every evening; Abeddin-Effendi's large house is crammed full of old treasures; and wives are entitled, Aziyadé says, to borrow from their master's stocks.

She will take all of it back when the dream comes to an end, while what belongs to me will be sold.

VII

Who will give me back my life in the Orient, my free life in the open air, my long leisurely walks, all the din and commotion of Stamboul?

To set off in the morning from Altmeidan with a mind to spend the night at Eyoub; rosary in hand, to tour round the mosques; to stop by every café, tomb, mausoleum, baths, and squares; to sip Turkish coffee from those tiny blue cups balanced on copper bases; to sit in the sun and let my mind drift gently with the hookah smoke; to chat with Dervishes and passers-by; to be an integral part of this canvas that's full of movement and light; to be free, without a care in the world, and to be unknown; and then to centre my thoughts on the beloved waiting for me at home at the end of the day.

My little friend Achmet makes a truly delightful walking companion; whether light-hearted or lost in thought, this ordinary young fellow, poetic to a fault, laughs all the time and would risk his life for me.

The picture darkens as we proceed into the old part of Stamboul by way of Eyoub, the holy quarter, and the large cemeteries there. We still catch glimpses of the blue water of Marmora, of its islands or of mountains in Asia, but there are very few people about and the houses are dismal;--the seal of age and mystery,--and various things outside bespeak wild tales of old Turkey.

More often than not it's nightfall when we get to Eyoub; we will have dined somewhere along the way in one of those little eating booths where Achmet personally checks that the ingredients are clean and supervises their preparation.

We light our lanterns before returning to the house,-- this house, so out of the way and peaceful, its very remoteness is one of its charms.

VIII

My friend Achmet is twenty, according to his old father, Ibrahim; twenty-two, according his old mother Fatma; Turks never know their age. Physically Achmet is something of a puzzle: he's small in stature, but of a powerful build; anyone who didn't know him could take his thin, sun burned face to signify a delicate constitution;--tiny aquiline nose, tiny mouth; small eyes full of gentle melancholy one moment,

twinkling with jollity and wit the next. Someone people really take to.

He's an unusual young man, joyful as a bird;--he entertains the most comical ideas and expresses them in a way that's entirely original; his views regarding honesty and honour are too exacting. He can't read and spends all his time on a horse. His heart is as open as his hand: he gives half of what he earns to the beggar women in the streets. The two horses he hires out to people constitute all that he possesses.

It took Achmet two days to find out who I really was, and he promised to guard the secret, which he alone knows, on condition that in future he be accepted as one of us. Gradually he has established himself as a friend, and has his place at our fireside. He adores Aziyade and is her servant knight; he is more concerned for her than she is for herself, and, with her interests in mind, he keeps a wary eye on me as if he were an old policeman.

'Take me on as your servant,' he said one day, 'in place of little Yousouf who is dirty and a thief. If you feel you want to give me something, you can give me what you give him. It'll be a bit droll, having me as a servant, but it means I'll live in and that will be fun.'

The following day I let Yousouf go and Achmet took over.

IX

When, a month later, in somewhat of a quandary, I offered Achmet—who is a byword for patience--two medjidies for pay, he flew into a purple rage and smashed two window panes, which he replaced the next day at his own expense. And that's the way the wages question was settled.

X

I can see him standing in my room one evening, tapping his foot.

'Sen tchok chéytan, Loti!... Anlamadum séni! (There's a lot of the devil in you, Loti! You're a real crafty one, Loti! I can't make you out at all!)'

He shook his arm angrily, so that his wide white sleeves waved from side to side; the bobbing of his small head caused the silk tassel on his fez to jiggle madly about.

It was all to do with the scheme that he and Aziyade had devised to get me to stay in Stamboul: he was to offer me half of what he owned--to wit, one of his horses--and I had laughingly declined. For doing so I was 'tchok chéytan' and deemed incomprehensible.

Ever since that evening I have been truly fond of him.

Dear little Aziyade! she had used up all her arguments and all her tears to keep me in Stamboul; the time set for my departure hovered like a black cloud over her happiness.

Then, when she was completely worn out: 'Benim djan senin, Loti... (My soul is yours, Loti. You are my God, my brother, my friend, my lover; when you've gone, it will be the end of Aziyade; her eyes will be closed, Aziyade will be dead.--Now do whatever you want; you're the one to know!')

'You're the one to know'--an untranslatable statement which amounts to saying: 'As for me I'm only a poor little girl who is at a loss to

understand you; I bow to your decision, and I shall

honour it. When you are gone I will go off into the mountains, and I will sing my song for you--Cheytanlar, djinler, Kaplanlar, duchmanlar, Arslandar, etc...

May devils, jinees, tigers, enemies stay far from my loved one...and I shall perish from hunger on the mountains singing my song for you.'

The song would then follow; every evening she sang it in a soft, lengthy monotone; the rhythm was strange, the pauses seemingly exaggerated, the final touches of sadness unmistakeably oriental.

When I am gone from Stamboul, when Aziyade is lost to me for ever, I will continue to hear her song in the night.

XI

to LOTI from his SISTER

Brightbury, December 1876.

'My dear brother,

'I have read and reread your letter! It is all I can expect to do for the present, and, like the Shulamite woman, gazing at her dead son, I am able to say: 'It is well!'

'Your poor heart is full of contradictions, as are all troubled hearts

that drift without a compass. You give cries of despair, you say

everything eludes you, you make impassioned appeals

to my affection, and then, when I, in return, give you impassioned assurances, I find that I and others from your former life are no longer carried in your thoughts, and that you are so glad to be in that Oriental nook of yours that your only wish would be that this Eden could last for ever. Yet here I am; I am constant and unchanging; and you will find me so when your sweet delusions are forgotten, giving place to other ones; perhaps later my being here will mean more to you than you think.

'Dear brother, you are mine, you are God's, you belong to us both. I have the feeling that one day, perhaps soon, you will regain courage, faith, and hope. You will see how sweet and delightful, how precious and beneficent is this 'fallacy', this thousand times blessed 'falsehood', that alone brings me life, as it will bring me death, with no fear or regret; this blessed 'falsehood' that has for centuries guided the world, creating martyrs and great nations; that changes mourning into joy, proclaiming: 'Love, Freedom, and Charity.

..........'

XII

Today, 10th December. My presentation at the Sultan's palace.

In the courts of the Dolma Bagtche Palace everything is white as snow, even underfoot: marble embankments, marble paving stones, marble steps. The Sultan's guards turned out in scarlet, the musicians dressed in sky blue laced with gold, and the footmen in apple green lined with capuchin yellow: all the colours stand out boldly against such unbelievable whiteness.

The pediments and cornices high on the palace walls serve as perches for families of gulls, divers, and storks.

Within the Palace all is splendour.

The Sultan's bodyguard, rigid as gilded mummies beneath their tall plumes, line the stairs. The guard's officers, looking rather like images of the renowned Aladdin, give their orders by sign.

The Sultan is solemn, pale, weary to the point of collapse.

The short formal presentation ended you bend almost to the ground, and giving deep salutes, you withdraw backwards.

Coffee is served in a large room overlooking the Bosphorus.

A kneeling servant will light you a two metre long tobacco pipe that has an amber mouthpiece, enriched with jewels and a bowl that rests on a silver tray.

The 'zarfs, (coffee-cup holders) are of chased silver set all over with large rose-cut diamonds and a host of other precious stones.

XIII

You could search throughout the whole of Islam and not find a husband more hapless than old Abeddin-Effendi; he's always away, always somewhere in Asia, he with four wives and not one of them over thirty. What is even more remarkable is that the four wives are as thick as thieves and guard one another's secrets as if they were their own.

Aziyade herself, by far the youngest and prettiest, is not hated enough by her elders for them to betray her.

Incidentally, she is their equal; she has undergone some ceremony or other which endowed her, like the others, with the titles of 'lady' and 'spouse.'

XIV

I asked Aziyade: 'What do you do when you're in your husband's house? How do you fill your time during those long days in the harem?'

'Me?' she replied, 'I get bored; I think of you, Loti; I look at your portrait; I feel your hair, or I play with various little odds and ends of yours that I take away with me to keep me company.'

To Aziyade, being in possession of anyone's portrait or lock of hair seemed the oddest thing ever; she would never have dreamt of doing so but for me; it was completely against Moslem teaching; something thought up by infidels; she was charmed by the practice, but there was always an element of fear present.

She must have really loved me to let me take some of her own hair; it made her tremble to think that she might die suddenly, before the hair had re-grown so that she would appear in the next world with a whole lock shorn clean off by an unbeliever.

I questioned her again: 'But before I arrived in Turkey, Aziyade, how did you spend your time?'

'Back then I was scarcely more than a child. When I saw you for the first time I had been in Abeddin's harem less than ten moons, and I hadn't yet grown tired of it. I would keep to my room, and sit on my

divan smoking cigarettes or hashish, or playing cards with my maid, or listening to strange stories from the black men's country that Kadidja tells really well.

'Fenzile-hanum taught me how to embroider, and then there were the visits we exchanged with ladies in other harems.

'We had our duty to our master, of course; and, lastly, we could be taken for drives. By rights each of us could have use of our husband's carriage for a day: but we would all rather go out together.

'Compared with how things are in other harems we get on very well.

'Fenzilé-hanum, who is very fond of me, is the eldest and the most important lady in the harem. Besme is the hot-tempered one and sometimes flies off the handle, but she's easy to calm down and its soon forgotten. Aïche is the nastiest of the four but she depends on us has to keep herself in check, because she is the one who is most shameful. She had the audacity once to bring her lover into her room! ...'

That had so often been my dream too, to get into Aziyade's room, just the once, simply to have some idea of where the girl I was fond of lived her life. We had given a good deal of thought to the plan, and had even consulted Fenzilé-hanum about it, but we didn't go through with it, and the more I've learnt about Turkish customs, the more I recognise that it had been a crazy idea.

'Ours,' concluded Aziyade, 'is considered everywhere to be a model harem, on account of our good-natured tolerance and agreement.'

'Some model, that!' I ventured. 'Are there many like it in Stamboul?'

It was Aiche-hanum--(pretty little madame!)--who was the source of the malady: in two years it has spread so fast that now the old man's house is no better than a hub of intrigue; all the servants have been corrupted. This large stern-looking house with its barred windows and doors has become a kind of conjuror's box: its secret doors and hidden staircases give the captive birds licence to leave, and off they fly in whatever direction takes their fancy.

XV

Christmas. A beautiful night. So clear, so star-studded, so cold.

At eleven o'clock I left the 'Deerhound' and went ashore at the foot of the old Foundoucli mosque, whose crescent gleamed in the moonlight.

Achmet was there waiting for me and we began our ascent through the quirky streets of the Turkish quarters of Pera.

Loud outbursts from the dogs. You'd think you were caught in some fantasic story book with illustrations by Gustave Doré.

I had been invited to a Christmas party up here in the European area of the town, a Christmas party like all those being held in every corner of the old country.

Ah! those Christmas nights of my childhood ... what tender memories I still cherish.

XVI

LOTI to PLUMKETT

Eyoub, 27 September 1876.

'Dear Plumkett,

'You'll know by now that Turkey, poor old Turkey, has proclaimed its constitution! I ask you, what is the world coming to? What century were we born in? A constitutional Sultan?--it plays havoc with all the notions concerning him that have been instilled in me.

'What has occurred has shocked people in Eyoub; all good Moslem folk are thinking that Allah has abandoned them, and that the Padishah is losing his mind. I, who regard all serious matters as something of a joke, especially politics, tell myself that simply from the point of view of its uniqueness Turkey will lose a great deal by the application of this new system.

'Today I was sitting with a few Dervishes in the funeral mosque of Suleiman the Magnificent. As we discussed the Koran we touched a little on politics, and we agreed that neither that great sovereign, who had his son Mustapha strangled before his eyes, nor his wife Roxelane, who gave us the 'Roxelane nose', would have the Constitution. A parliamentary government will be the ruin of Turkey, of that there is no doubt. '

XVII

Stamboul, 27 September.

7 Zi-il-iddjé 1293 of the Hegira.

Waiting for a shower to finish I went into a Turkish cafe near the Bayazid mosque.

The only people inside were a few turbaned old men (hadj-baba) with white beards who sat immersed in newspapers or gazing through the smoke-smeared windows at the passers-by rushing past in the rain. Turkish ladies, caught in the downpour, sped along as fast as their heelless slippers and wood-soled sandals would let them. There was great confusion in the street and much jostling; the rain was coming down in torrents.

I looked at the old men about me: the way they dressed indicated meticulous adherence to the fashions of the good old days; everything about them was 'eski', from their great silver-rimmed spectacles to the contours of their aged profiles. 'Eski', a word pronounced reverently, meaning 'antique', is equally applicable in Turkey to old customs, old styles of dress, or old fabrics. The Turks have a love of the past, of fixity and stagnation.

Suddenly there came the sound of cannon fire, a gun salute from the Seraskierat. The old men exchanged knowing signs and ironical smiles.

'Hail to the Constitution of Midhat-pacha,' said one of them, executing a mocking bow.

'Deputies! A charter!' muttered another old man in green turban, 'The khalifs of old had no need at all of the people's representative.

'Voï, voï, voï, Allah!.. nor did our wives need to go around in gauze veils; and the faithful said their prayers more regularly; and the Muscovites* were not so insolent!'

The military salvo announced to the Moslems that the Padishah vouchsafed to them a constitution that was more extensive and liberal

than any of the European ones; and yet this gift from their sovereign got a very cold reception from these old Turks.

This event, which Ignatieff had done his damnest to delay, had long been expected; you could say it marked the day when the Porte and the Tsar tacitly declared war against eachother and when the Sultan zealously pushed ahead with his arms provision.

It was half past seven by Turkish time (about midday). The proclaiming of the new Constitution was taking place at Top-Kapou (Sublime Portal), and, hurtling through the heavy rain, that is where I headed.

The Viziers, the Pashas, the generals, all the government officials, all those in high places, every one in full fig, their gilt shining, were crowded into the great Top-Kapou square together with the court bands.

The sky was dark and stormy; heavy falls of rain and hail swamped the whole assembly; it was as if the new charter were being delivered to

the people beneath a cataract; the old crenelated walls of the Seraglio curtaining the scene seemed greatly amazed to hear such subversive phrases in the very heart of Stamboul.

Shouts, cheers, and fanfares concluded this remarkable ceremony; then the whole crowd, soaked to the skin, dispersed amid noise and excitement.

Simultaneously, at the other end of Constantinople, members of the International Conference were in session in the Admiralty Palace.

The timing was deliberate: the salvos were to be sounded in the middle of Safvet Pasha's speech to the plenipotentiaries, adding effect to his peroration.

XVIII

'The East! The East! Poets, what do you see?
Turn your eyes and minds eastward!
'Alas!' they cried in voices long since still,
'We can see well a mysterious daylight out there!
. . . .
Perhaps we take the twilight to be a dawn'
. . . .

(VICTOR HUGO, 'Songs of Twilight'.)

I shall never forget gazing around me in the great square of the Seraskierat, that immense esplanade on the heights of central Stamboul, and letting my eyes wander into the distance as far as the mountains of Asia. The Arab porticos, the high tower whose weird shapes were illuminated, as on nights of important festivals. The torrent of rain that had fallen that day had turned the square into a veritable lake that reflected all the rows of lights; rising into the sky around the vast horizon were the domes of the mosques and of the pointed minarets, their long stems each holding aloft a crown of ethereal brilliance.

In the square a deathly silence reigned; not a soul was to be seen.

Across the clear sky, swept by a wind that one could only imagine, ran two bands of black clouds, and above them the moon had just affixed its bluish crescent. It was one of those indifferent touches nature seems

to make when when some momentous event in the history of nations is about to occur.

There came a noise of many feet and voices; a band of softas, carrying banners and lanterns came into the square through the central

porticoes, shouting 'Long live the Sultan! Long live Midhat-pasha! Hurrah for the Constitution! Hurrah for the war!' It was as if the men were drunk with the notion of being free; but a few old Turks, remembering the past, shrugged their shoulders as they

watched droves of exaltant people moving around the square.

'Let us go and salute Midhat Pasha,' cried some of the softas.

They went off to the left to make their way by small secluded streets to the modest house where dwelt the great Vizir, such a powerful figure then, who only a few weeks later was to be forced into exile.

Softas numbering about two thousand took themselves to the great mosque (the Suleiman) to offer prayers, and from there they crossed the Golden Horn and continued to Dolma-Bagtche in order to acclaim Abdul Hamid.

Outside the palace railings deputations from all associations instinctively joined company with a huge conglomeration of people with the shared purpose of giving their constitutional monarch an enthusiastic ovation.

The same groups returned to Stamboul along Great Pera Road, giving ovations as they went, one for Lord Salisbury (who was soon to become so unpopular), one for the British embassy, and one for the French.

'Our ancestors,' said the hojas haranguing the crowd, 'our ancestors, who numbered no more than a few hundred men, conquered this land four centuries ago! Are we, who are several hundreds of thousands strong, going to let it be invaded by the foreigner? Let all of us, Moslem and Christian together, die for our Ottoman fatherland, rather than accept humiliating terms ...'

XIX

The mosque of Sultan Mehmed-Fatih (Mehmed the Conqueror) often sees us, Achmet and myself, sitting outside its great grey-stone particoes, both of us leant back in the sunshine without a care in the world, dreaming some vague dream or other, untranslatable into any human language.

The square of Mehmed-Fatih, high above old Stamboul, occupies large open spaces, where men in cashmere caftans and wide white turbans can be seen strolling about. At the centre of the square stands the mosque, one of the largest and most venerated in Constantinople.

The immense square is surrounded by strange-looking walls, which are topped by lines of stone built domes,arranged like rows of beehives; they are where the softas live, and infidels are not allowed to enter them.

This quarter is the centre of a daily life that is unmistakably oriental. Camels cross it at their leisurely pace, their bells tinkling monotonously; dervishes come to sit there and converse about holy matters, and as yet there is not a trace of western influence.

XX

Not far from the square is a dismal, unfrequented street where grass and moss are left unchecked. That's where Aziyade's home is; and for me that is the secret attraction of this square. Throughout those long hours during the day, when I can't be with her, I'm there, not too far away, I'm where no one knows me, where I'm safe from any suspicion.

XXI

Aziyade is lapsing into silence more often, and there is a deeper sadness in her eyes.

'What's the matter with you, Loti?' she asked, 'Why are you always gloomy? It's I who should be gloomy, because I will die when you are gone.'

And she fixed her eyes on mine with so much penetration and so much tenacity that I turned my head away from her.

'I gloomy, my darling!' I said, 'I lack nothing when you're here. I'm happier than a king.'

'Indeed you are; is there anyone more loved than you, Loti? Is there anyone you could be really envious of? The Sultan, perhaps?'

She's right. The Sultan, who, in the eyes of the Ottomans, cannot fail to enjoy the most pleasure a man can have, is someone I don't envy. He is grown old and weary; furthermore he's now a 'constitutionalist'.

'I think, Aziyade,' I said, 'I think the Padisha would give everything he owned--even his great emerald, which is as big as my hand, even his charter and his Parliament —if he could have my liberty and my youth.'

I'd wanted to say: '...and have you too!...' but doubtless the Padishah would have little regard for any woman, however charming. Besides, the

last thing I wanted to do was to utter some comic opera cliché. I was certainly dressed for the part. I didn't like what the mirror told me: I'd made myself look like a young tenor ready to deliver a piece from Auber.

It so happens that there times when I no longer manage to take my role as a Turk seriously; Loti's ears don't quite fit beneath Arif's turban, and like a fool I relapse into my old self, and am left feeling unbearably dull.

XXII

I could never accept all that being an officer and gentleman entailed. I never encountered anyone either brilliant or grand enough for me. I had a great deal of contempt for my peers and selected my friends from among the more sophisticated of them. Here I've become a man of the people, an ordinary citizen of Eyoub; I make the best of the humble life of the boatmen and fisherfolk, even their company and their pleasures.

When, in the evening I enter Suleiman's cafe with Samuel and Achmet, the circle round the fire is widened to make room. I shake hands with everyone and sit down to listen to someone telling winter night tales (long stories about djinns and genies that can last a week). Hours pass and I find I've grown neither weary nor sad; I feel at ease with these people and not at all disoriented.

Arif and Loti being two entirely different persons, on the day the 'Deerhound' sails, Arif need only remain in his house, where no one would think of looking for him; but Loti would have disappeared and disappeared for ever.

Now and then this idea, which came from Aziyade re-enters my mind in remarkably appealing guises.

To remain with her, not in Stamboul any more, but in some Turkish village by the sea; to lead the life of ordinary folk, out in the sun and fresh air; to take every day as it comes, free from creditors and worries about the future! I'm more suited to that kind of life than I

am to the one I have. I loathe all work where I can't use my body and muscles; I loathe anything to do with science; I hate all the conventional duties and social obligations that prevail in our western countries.

To be a boatman in a gold-embroidered jacket, some place in southern Turkey, where the sky is always clear and the sun is always hot!

It's possible, after all, and I would be less unhappy than I would be elsewhere.

'I swear to you, Aziyade, I would give up everything without a moment's regret--my social standing, my name, and my country. My friends...I have no friends and I don't care for any! But, you see, I have an old mother.'

Aziyade no longer asks me outright to stay with her, though perhaps she's understood it might not be entirely impossible; but her intuition tells her what having an old mother means, tells this poor little girl who never had one; and all the ideas she has regarding generosity and sacrifice are of greater worth in Aziyade than in others because they came to her naturally; no one troubled themselves to impart them to her.

XXIII

From PLUMKETT to LOTI
Liverpool, 1876

'My dear Loti,

'Figaro was a man of genius: he laughed so often, he never had time to cry.-- His guiding principle is the best one of all, and I'm so

convinced that I endeavour to put it into practice and more or less succeed.

'Regrettably, I find it difficult to remain the same person for too long. Too often Figaro's cheerfulness deserts me, at which point Jeremiah, the prophet of woe, or the majestically despairing David, upon whom the celestial hand lies heavy, seizes me and takes control of me. I don't speak--I shout, I howl. I don't write--I would only break my pen and knock over my ink bottle. I pace up and down, shaking my fist at an imaginary being, at the best scapegoat I can think of, laying all my sorrows upon it. I'm extravagant in every possible way: behind my locked door I indulge in the most insane acts, and afterwards, more wearied than relieved, I calm down and become reasonable again.

'You're going to tell me yet again that I'm a weird sort of chap, a madman, and goodness knows what else. To which I'll reply: 'Yes, but I'm not half as crazy as you imagine—not half as much as you, for instance.'

'Before passing judgement you must needs have closer acquaintance with me, and understand me a little better and know the sort of circumstances that could turn a fellow, who was sensible from the day he was born, into the odd individual I am today. You see, we are the product of two factors, the first being our inherited natures, which we bring with us onto life's stage; the second being all the circumstances that alter and fashion us as if we were made of a plastic substance that receives and retains the imprint of everything that has touched it.-- In my case the circumstances have all been of a painful kind. I was taught (forgive my time-worn phrase) at the school of misfortune. All the knowledge I've gained came at a price; that is why I sometimes express myself rather trenchantly. If I occasionally appear dogmatic, it's because I'm claiming, on the grounds of my suffering, to know more

on various subjects than those who have suffered less, and to be able to speak more authoratively than those who don't have full knowledge of the facts.

'As I see it, this world is barren of hope, and I do not have the consolation of those whose ardent faith makes them fit and strong for life's struggles, and able to trust in the supreme justice of the creator.

'Nevertheless, mark you, I live without blaspheming.

'So, beset continually by hurts and disappointments, have I been able, then, to hang on to the illusions, the enthusiasm, and the spirited freshness of my youth? No, as you very well know; I've renounced such delights as are appropriate to my years, simply because they don't appeal to me any more; I no longer look or behave like a young man, and from now on I shall be living a life that has neither purpose nor hope... But does this mean that I'm reduced to the same state as yourself, sick of everything, denying all that is good, denying virtue and friendship, denying everything that can raise us above the level of the beast?

'My friend, let's be clear about this; on all those points my thinking is totally different from yours. I have to say that despite all that life has taught me (and may you never acquire anything resembling it, as it comes at too high a price!) I still beieve in all those things, and in much else besides.

'When I was in London, George showed me the letter he'd just received from you.

'You begin your letter very teasingly with a lively and detailed account in a Turkish-style rendering of one of your passing infatuations. We, George and I, follow you all through your shadowy meanderings as if we're in some huge, oriental ant-hill. We are left, mouths agape, viewing the pictures you paint for us. Those three daggers of yours bring into my mind Achilles' shield, celebrated in song with such meticulous care by Homer! And then finally, perhaps because you got a speck of dust in your eye, or because your lamp began to smoke as you were finishing your letter, perhaps on account of something more trifling still, you wind up by pitching at us a string of commonplaces that belong to the last century! I actually believe that the platitudes of the lay brothers are worth far more than those of the materialists who will end up annihilating everything that exists. In the eighteenth century these were accepted materialist ideas: God was only an assertion; morality meant convenience; society was one vast entity ready for the next clever man to exploit it. A good many people were seduced by the novelty of all this and by the approval it gave to the most immoral actions. Blessed epoch when restraints were gone; when you could do as you pleased; when you could laugh at everything, even things that were not in the least funny, until the time came when so many heads had dropped from the Revolution's blade that those who had kept theirs began to reflect. There then followed a period of transition during which one saw appear a generation ravaged by moral decay, afflicted with constitutional sentimentality, lamenting the past it had no connection with, rueing the present it didn't understand, doubtful of the future it couldn't imagine. A generation of romantics, a generation of young folk spending their days laughing, crying, praying, blaspheming, trying every change of tone in their

vapid lament, until one day they reach the point when they blow their brains out.

'These days, my friend, people are much more rational, much more practical: on the way to becoming a man you make haste to model yourself on the type of man—or distinctive animal, if you prefer--that you have selected. The whole set of opinions or prejudices that are typical, you make your own; you have chosen a certain social group and you share its ideas. By these means you develop a distinct cast of mind, or, if you will, a kind of stupidity that is fitting for and specific to this world in which you live. Your associates understand you, you understand them; consequently you enter into intimate communion with them and become literally a member of their body.

'Banker, engineer, bureaucrat, grocer, soldier . . . whatever it be. But at least you're something, and you have something that you do; your head is firmly on your shoulders and you're not off chasing endless dreams. You're sure of yourself; your line of conduct is clearly marked out by the duties you are professionally bound to fulfill. There are handbooks that cover any queries you may have in regard to philosophy, to religion, to politics, so don't make a burden of such small matters. Civilisation absorbs you; the thousand and one wheels of the huge social machine enmesh you and jig you about all over the place, leaving you stupefied; the children you have will be as stupefied as yourself. Then finally, having received the rites of Holy Church, you die; your coffin is awash with holy water; in faint light of candles around your bier Latin hymns are sung in fauxbordon harmony; those who had been accustomed to seeing you will miss you if you have led a good life, several will even shed genuine tears over you. When all is done, your heirs inherit what was yours.

'And so the world goes on.

'None of this, my friend, obscures the fact that there are on this earth some really good people, who are thoroughly decent by nature, who do good because they derive personal satisfaction from doing so: they neither steal nor take life even when they could do so with impunity, because they have a conscience which is a perpetual control against actions which their passions could drive them to commit; people capable of loving, of giving of themselves body and soul; priests who believe in God and who practise Christian charity; doctors, defying epidemics in order to save the lives of a few poor sick patients; sisters of mercy tending to wounded soldiers on the battlefield; bankers who can be trusted with your money; friends who would give you half of theirs; people—myself, for example, to seek no further—who would perhaps, despite all your blasphemies, be capable of offering you boundless affection and devotion.

'So no more of those vagaries of a sick child. They result from dreaming instead of thinking, from addiction instead of reason.

'You slander yourself when you talk as you do. If I were to tell you that I agreed with everything you wrote at the end of your letter, and that I believe your self-portrayal to be a true one, you would write to me on the spot to protest, informing me that you didn't mean a single word of your appalling profession of faith; that it was no more than the bravado of a heart more loving than other people's; only the painful effort a sensitive plant makes when wrung with sorrow.

'No, no, my friend, I do not believe you, nor do you believe yourself. You are good and affectionate, you are sensitive and refined. Only, you are in pain. And therefore I forgive you, and shall remain a living protest against your denial of all that stands for friendship, selflessness, devotion.

'It's your vanity that's denying all of those things, not you; your wounded pride causes you to hide your treasures and just for the sake of it to flaunt 'the sham human being that your arrogance and boredom have created.'

'PLUMKETT.'

XXIV

LOTI to WILLIAM BROWN

Eyoub, December 1876.

'My dear Brown.

'This is to remind you that I'm still in the land of the living. I go by the name of Arif-Effendi, and I would be delighted to know if I can hope for signs of life from you.

'Disembarking on the Stamboul side of Constantinople; you thread your way through four kilometres of bazaars and mosques; you come to the sacred part of the city, Eyoub, where the children make your bizarre headgear a target for their pebbles. You ask directions for Kourou-Tchechmeh Street and will have your answer immediately; at the end of this street, beneath some almond trees, there's a marble fountain, my house is right next to it. I live here with Aziyade, the girl from Salonica I used to tell you about, the girl I'm not very far from loving. I'm almost happy living here, forgetful of the past and all the its heartlessness.

'I'll spare you the details of what brought me to this eastern hideaway and how I came to adopt for a while the language and customs of Turkey—even its beautiful silk and gold garments.

'This is what it's like here tonight, 30th December: clear sky, cold, moonlight—the monotonous chant of the dervishes drifts into the night air, a familiar sound that rings daily in my ears. My cat Kedi-bey and my servant

Achmet, the one in the arms of the other, have retired for the night to the room they share.

'Aziyade, as befits a daughter of the East, is seated on a stack of rugs and cushions, busy tinting her nails a reddish orange, an operation of high importance.

'As for me, I find myself thinking of you, of the life we led back in London, and all the foolish things we did,--and I'm writing to you in the hope of receiving a reply from you.

'I'm not a proper Moslem, though you might have supposed so from the way I began this letter; quite simply, I lead a double life, but I'm still officially, albeit the least often possible, M. Loti, the naval lieutenant.

'Don't torture yourself copying my address in Turkish, use my real name and send it by way of the 'Deerhound' or the British Embassy.'

XXV

Stamboul, 1st January 1877.

The year 77 came in with a dazzling, springlike day.

Obliged by what little condescension I still had in the matter of Western customs, I spent the day attending to a number of calls in the Pera community; I rode back home to Eyoub that evening by way of the cemeteries and Kassim-Pacha.

I passed the terrible Ignatieff in his brougham; he was hastening back from the Conference under a heavy escort

118

of Croats, all of them in his pay; a moment later Lord Salisbury and the English Ambassador followed, both of them greatly agitated: heated words had been exchanged at the meeting, and things were as bad as they could be.

With the energy born of despair the hapless Turks rejected the terms imposed upon them; for their pains, they could find themselves without legal safeguards.

That being the case, the ambassadors would all quit the meeting together, shouting 'Every man for himself!' to the Europeans. One could picture terrible things ahead, huge disarray and much bloodshed.

Would that we be spared this catastrophe!...

It would mean my leaving Eyoub, perhaps even tomorrow, never to return ...

XXVI

One glorious evening we made our way down the steep roadway towards Oun-Capan.

Stamboul seemed different; hodjas in every minaret were chanting unfamiliar prayers to a strange music; the sound of high-pitched voices emanating from such heights and at such an unusual time of night preyed on the imagination; all the mussulmans stood in groups outside their doors, their faces turned to something in the sky that they feared.

Achmet followed their gaze and seized my hand, terrified: the moon, which a few moments ago we had seen shining so brilliantly on the dome of St.Sophia, was no more to be seen in the immensity of space; all there was was a reddish mark, dull like blood.

There is nothing that grips my attention more than signs in the skies, and my first impression, one that registered almost instantly, was one of terror. Having ignored the calendar for ages, I was totally unprepared for what I saw taking place.

Achmet explained to me just how grave and ominous it was: according to Turkish belief, the moon was there and then at grips with a dragon that sought to devour it. It could be rescued, however, by pleading with Allah and by firing bullets at the monster.

Requisite prayers were, indeed, being said in all the mosques, and shooting had begun in Stamboul. From every window, from every rooftop, shots were being fired at the moon in the hope of securing a happy ending for that terrifying phenomenon.

At Phanar, we took a caique to get us home; but our journey was halted. Halfway across the Golden Horn, the police cutter barred our way: on the night of an eclipse it's an offence to go out in a caique. That's as may be, but we couldn't sleep in the street. Adopting a very high tone we parleyed and argued with our official interceptors, and once more our audacity won through.

We arrived at the house and found Aziyade waiting for us, terrified out of her wits.

The dogs were howling pitiably at the moon, which only complicated the situation further.

Wearing mystical expressions, Achmet and Aziyade informed me that the reason the dogs were howling was because they were asking God to give them the mysterious bread which is fed to them on certain solemn occasions—and which were invisible to human eyes.

Despite the shooting the eclipse progressed; the entire disc even took on a hue of an extraordinarily intense red, a colouration due to atmospheric factors.

I made an attempt to explain the phenomenon in the classic way schools do, using a candle, an orange, and a mirror.

I exhausted all my reasonings and assertions, and my pupils hadn't understood a thing; my hypothesis that the earth is round was rejected right at the start. Refusing out of hand to take me seriously, Aziyade, in her dignified manner, sat down. I tried looking like a teacher. What a hideous sight! Then I got the giggles. I ate the orange, and abandoned my demonstration ...

Anyway, of what benefit was all this stupid science to them? And why try to rid them of superstions which, in fact, only added to their charm?

So there we were, the three of us, joining the others and taking pot shots through the window at the moon, which, high above, continued to glow blood red, surrounded by shimmering stars in the most radiant of skies!

XXVII

About eleven o'clock Achmet woke us up to tell us that the treatment had worked; the moon was 'eyu yapilmich' (cured).

Sure enough, the moon, completely restored to health, shone like a fantasic blue lamp in the beautiful Eastern sky.

XXVIII

The old woman they call 'My mother Behidje' is a
very extraordinary person. Now in her eighties and
crippled, this daughter of one pasha and widow of
another, is more Moslem than the Koran and stricter than
Moslem Canon Law.

The late Chefket-Daoub-pasha, Behidje-hanum's
husband and one of Sultan Mahmoud's favourites, was
implicated in the massacre of the Janissaries. That was at
a time when Behidje acted as one of her husband's
advisers and she urged him on with all her might.

The old lady lives up on the heights of Taxim in the
Turkish quarter of

Djianghir in a street that goes straight up the hillside.
Although the house she lives in already overhangs
precipices, the two 'shaknisirs' projecting from it have
been carefully latticed in strips of oak.

From there you have a bird's eye view of the different
areas of Foundoucli, the Dolma-Bagtche and the
Tcheraghan palaces, Serail Point, the Bosphorus, the
Deerhound (looking like a nutshell lying on top of a blue
cloth) and then across to Scutari and the whole coast of
Asia Minor.

Behidje-hanum spends her days in this observatory,
outstretched on a reclining chair, and Aziyade often sits
at her feet,--Aziyade, watching for the slightest sign from
her aged friend, and hanging on her every word as if they
were divine decrees from an oracle.

It's not at all what you'd expect, this closeness between the humble young woman and the old honoured wife, proud and unbending, of noble stock, and living in a grand mansion.

I know Behidje-hanum only by repute: infidels are never allowed to enter her abode.

She is still beautiful, so Aziyade maintains, despite her eighty years: 'beautiful as a lovely winter's night'.

And every time Aziyade advances some new idea, or some elegantly profound concept concerning matters she couldn't possibly know about, and I ask her: 'Who did you learn that from, darling?' Aziyade will reply: 'From my mother Behidje.'

'My mother' and 'my father' are titles of respect that they use in Turkey when they're speaking of persons well on in years, even when such persons are just a name to you.

Behidje-hanum is most definitely no mother to Aziyade; or if she is, she's at best a reckless mother who thinks nothing of taking her child's imagination to dizzying heights. First and foremost it's in matters religious where she works on Aziyade so relentlessly and effectively that the poor little waif often sheds very bitter tears over her love for an unbeliever.

She gets Aziyade all stirred up as well with tales of yore, telling her lengthy romantic stories, imbuing them with fire and spirit--the same stories that are recounted to me in the evening by the moist lips of my beloved.

I hear of the stupendous adventures of Genghis the Great, or ancient heroes of the desert, and then there are the fairy tales, both Persian and Tartar, that tell of young princesses, who, though persecuted by jinees, accomplish marvels of fidelity and courage.

So, whenever Aziyade returns here in the evening with her imagination more over-excited than usual, I can safely say to her:

'My darling girl, you've spent the day sitting at the feet of your mother Behidje!'

XXIX

January 1877.

For the past week I've been at Buyukdere in the upper Bosphorus, where it joins the Black Sea. The 'Deerhound' is anchored close to the great Turkish ironclads that are stationed here as watch dogs, on the alert for any Russian intrusion. Moving the 'Deerhound' to a new berth puts me at a fair distance from Stamboul, but it coincides with old man Abeddin's visit to his Stamboul residence; so it's all for the best, and our separation relieves us of the need to be careful.

It's cold and rainy; I spend the daytime roaming Belgrade Forest; its leafy paths take me back to the happy days of my childhood.

The ancient oaks, the holly trees, the moss and the ferns could almost pass for the ones in Yorkshire. Were it not for the added presence of bears, I could believe myself back in the dear old woods at home.

XXX

Samuel has a fear of 'kedis' (cats). By day the 'kedis' put weird ideas into his head; he can't look at them without laughing. At night he becomes very respectful towards them and keeps them at a distance.

I was getting ready to go to an embassy ball when Samuel, who had just left me to go to bed, suddenly returned and knocked at my door. He

sounded alarmed: 'Bir madame kedi (A 'lady' cat; read--'female' or 'pussy')--qui portate ses piccolos dormir com Samuel' (who has brought her little ones to sleep with Samuel)!

And with impassive seriousness he continued from the wings:

'In my family, we believe that anybody who disturbs cats, will die within that same month! Monsieur Loti, what am I to do?'

When I'd finished dressing I decided to lend my friend a hand, and I went unto his room.

A 'lady kedi' had indeed stationed herself on Samuel's pillow, right in the middle of it. She was considerably stout and had a beautiful yellow coat. She looked dignified and triumphant, sitting on her 'unmentionable', gazing in turn at a motionless Samuel and her little ones who were frolicking on the blanket.

Samuel, sitting in a corner, scarcely able to keep his eyes open, gazed on this family scene with resignation and dismay; he was waiting for me to come to his rescue.

I had not yet made the acquaintance of Madame Kedi, however, she raised no objection to my taking hold of her, lifting her onto my shoulder, and putting her and her offspring outside; after which Samuel, when he'd given his blanket a good shake, made it clear that he wanted to turn in.

I had absolutely no reason to return to the house that night, but, as things worked out, I got back at two in the morning.

Samuel had thrown the window of his room wide open, put up clothes lines, and hung out all his bedclothes, aiming to let the fresh air rid them of any lingering cat odour. He himself lay curled up in my bed sleeping the sleep of youth and innocence. All very well for him ...

Next morning we learned that this Madame Kedi was given to wandering and was the darling pet of an old Jew, a shoe mender, who lived in the neighbourhood.

XXXI

It was the Greek Christmas; the old neighbourhood of Phanar was in festive spirit. Bands of children went through the streets carrying

lanterns and paper spinners of all shapes and colours; they banged on everybody's door with all their might and delivered awfully bad serenades to the accompaniment of a drum.

Achmet, who was accompanying me, exhibited his great contempt for these infidel rejoicings.

Old Phanar, even with all this noise going on around it, couldn't quite shed its normal dreariness.

Yet you would see all the small, outdated, time-worn byzantine doors being opened; and, framed by the massive doorways, you would see young women appear, dressed like Parisians, and throw copper piastres to the musicians.

Arriving at Galata we could see that things there were far worse; there's not a country on earth where you'd ever come across a more discordant din nor a more wretched spectacle.

We could hardly credit the swarming conglomeration of different nationalities. By far the greater number were Greek; these loathsome inhabitants, issuing from every backstreet brothel, every bar, every

tavern, flocked into tight masses. Impossible to imagine the number of drunken men and women there, or the amount of wine-throated yelling and sickening screams.

And several good Moslems were there too. They'd come to enjoy a quiet laugh at the infidels' expense, and to see how these Levantine Christians--whose fate, so poignantly spoken of in speeches, had moved the hearts of Europe--celebrated the birth of their Prophet.

Ever since the day the Constitution conferred on them the unmerited title of citizen, these men--every one of them terrified at the thought of having to go and fight as Turks--had devoted themselves wholeheartedly to wine and song.

XXXII

I remember the night when the 'bay-kouch' (owl), followed our caique on the Golden Horn.

It was a cold January night. A freezing mist, falling like a fine rain on our heads, caused all the massive outlines of Stamboul to run into one. Achmet and I took turns at the oars as we headed in the direction of Eyoub.

Nearing the Phanar quayside we cautiously threaded our way in the dark through all the posts, the wrecks, and the countless caiques lying aground on the mud.

We reached a spot at the foot of the old walls of the Byzantine quarter of Constantinople, a place no human being would visit at that late hour. And yet, two women were sitting there, huddled together, waiting, two white-headed shadows tucked away in a particular dark corner beneath the balcony of a ruined house ... They were Aziyade, and the old, faithful Kadidja.

As soon as Aziyade was settled in our boat we took off again.

There was still a fair distance to go to the Eyoub quay. Every now and then the occasional light coming from a Greek house would cast a yellow trail into the murky water; for the rest of the time darkness reigned.

As we passed along the front of an age-old house that was covered in ironwork, we caught the sound of an orchestra and of a ball in progress. It was one of those large mansions, dark on the outside, sumptuous within, where Phanar's older Greek families hide their wealth, their diamonds, and their Paris clothes.

... Then the sound of the festivity died away in the mist and we re-entered a world of silence and darkness.

A bird insisted on flying round our caique above our heads, first one way and then the other.

'Bou fena', (not good)! said Achmet, shaking his head.

'Bay-Kouch mi?' Aziyade asked him, hidden by her hood and her tightly wrapped clothing. (It's not an owl, is it?)

When it was a matter concerning their superstitions or their beliefs it was customary for the two of them to reserve discussion for themselves and to disregard me completely.

After a pause she took hold of my hand. 'Bou tchok fena Loti', she said,'ammâ sen ... bilmezsen!' (It's very bad, Loti, but you ..., you don't know ...)

But it was odd, to say the least, seeing this creature going round and round on a winter's night, and it followed us relentlessly for an hour or more, the time it took us to go from the waterfront at Phanar to the one at Eyoub.

That night there was a dreadful current running in the Golden Horn; the freezing drizzle never ceased; our lantern had gone out, which meant we ran the risk of being arrested by the 'bachibozouks' on patrol, and that would have been a catastrophy for all three of us.

Near the Balata ferry crossing, we met caiques full of 'iaodis' (Jews) who, at this point, occupy both banks, at Balata and opposite at Pri-Pacha. They would have been paying evening visits, or returning from the great synagogue. At night this is the only place on the Golden Horn where there's any activity.

As they went by they were singing a plaintive song in their own language. The 'bay-kouch' continued to execute its turns above our heads, and Aziyade was weeping with cold and fear.

How glad we all were when, noiselessly and in pitch dark, we tied up our caique at Eyoub steps! Such a relief to be jumping from plank to plank over the mud (we were so familiar with these planks we could have done it blindfold); such a relief to be crossing the small, deserted square; to be gently turning the locks and withdrawing the bolts and to be closing them firmly behind all three of us; to be making a quick tour of the dim, ground-floor rooms, the space under the stairs, the kitchen and the inside of the oven; to be taking off our mud-filled shoes and our wet clothes and going barefoot up the carpeted steps; to be wishing good-night to Achmet who was retiring to his room; to be going into our room and closing the door and locking it again; to be letting down the red and white Arab door curtain and sitting on the piles of thick rugs in front of the copper brazier that has been smouldering since morning and is giving off a gentle heat, and the scents of 'pastilles du serail' and rosewater. It meant that for the next twenty-four hours, at least, we could enjoy both our safe return and the sheer bliss of being together.

But the 'bay-kouch' had followed us and started hooting beneath our windows.

And Aziyade, completely exhausted, fell asleep with the dismal sound in her ears and hot tears running down her cheeks.

'Our Madame', as Samuel and Achmet called her, was an oldish woman, no stranger to mischief, but likable. She'd been all over Europe and had tried her hand at everything. Bizum Madame was the phrase they used. Bizum Madame could speak every language, and she kept a sleazy cafe in the Galata quarter.

Bizum Madame's cafe looked out onto a busy main thoroughfare. The interior was huge and went a long way back; there was a back door opening onto a blind alley,and connecting with the Galata quays. The alley was used by those leaving places of ill repute, of which there were several. The cafe was, first and foremost, the meeting place for a number of Italian and Maltese merchant seamen who were thought to be trading in stolen and contraband goods. A lot of negotiating went on and to go there at night it was wise to carry a revolver.

Bizum Madame was very fond of all three of us, Samuel, Achmet, and me. Usually it was she who prepared the meals for my two friends, their own businesses often keeping them in this part of the city; Bizum Madame constantly mothered us.

On the first floor there was a small room with a chest in it. It was handy for changing 'between scenes'. I would enter by the main door, dressed as a European, and then leave by the door to the alleyway, a Turk.

Bizum Madame was Italian.

Eyoub, 20 January.

Yesterday the huge farce put on by the members of the International Conference simply petered out. The whole thing having flopped, their Excellencies are gone, the ambassadors are packing their bags, and, lo and behold, the Turks are left without any safeguards at all.

So bon voyage to all those people! Luckily for us, we're staying. In Eyoub everyone is very calm and sufficiently resolute. Of an evening in every Turkish cafe, no matter how lowly, one witnesses an instinctive coming together of rich and poor, of pashas and common folk ...

(O Equality! unknown to our democratic nation and and our European republics!) There's bound to be some knowledgeable person there who can fathom for those present the arcane texts of the daily newspapers; every individual will be listening in silence with conviction written across their brows. A world away from the noisy disputes, fuelled by ale and absinthe, that one finds in roped-off bars. In Eyoub they take politics very seriously.

One mustn't despair of a people that is principled and straightforward, a people that has kept true to so many of its beliefs,

XXXV

Today, 22nd January, the ministers and leading dignitaries of the Empire gathered in solemn session at the Sublime Portal and decided unanimously to reject the European proposals, beneath which they discerned the marks of Holy Russia's claw. Very soon, from all corners of the Empire, there arrived messages praising and congratulating the men who had made this desperate resolution.

National feeling ran high in this assembly; never before had there been such scenes: Christians seated side by side with Moslems; Armenian prelates alongside Dervishes and the Sheik-ul-islam. In this assembly for the very first time one heard from Moslem lips the incredidle words:'Our Christian brothers.'

In the face of a common peril a tremendous spirit of brotherhood and unity arose and was indeed bringing together the different religious communities in the Ottoman Empire. The Prelate of the Armenian Catholic Church spoke these remarkably warlike words before the assembly:

'Effendis!

'For five centuries the ashes of the forefathers of each and every one of us have lain peacefully in the soil of our homeland. The first and foremost duty is to defend this soil which has been vouchsafed to us as our inheritance. Death is a reality; it is the law of nature. History shows us great nation states, which, one after the other, have appeared, only to later disappear from the world stage. If, then, the decrees of Providence have numbered the days of our homeland we can do no more than bow to its ending. But it is one thing to fade away ignobly, and another to end in glory. If we are to perish, let us not forego the honour of meeting the fatal bullet full in the chest, not in the back; at least then our country's name will figure gloriously in history. Only a short time ago we were nothing but a listless body; the charter that has been granted to us revitalises this body and strengthens it.-- Today for the first time we have been invited to this council; for this we tender thanks to His Majesty the Sultan as well as to the ministers of the Sublime Portal. Henceforth let the religious question be confined to one's conscience; let the Moslem go to his mosque and the Christian to his church; but, keeping our eyes squarely on our common cause, and on the national enemy, let us all be united and remain so!'

XXXVI

Aziyade's loyalty to the little, open, heelless, yellow morocco babouches, as worn by pious Moslem ladies, never wavered; she went through a good three pairs a week; there were always spare ones left lying about in every corner of the house. She would write her name inside each one, using the excuse that Achmet or I might take them from her.

The ones that were past wearing faced a distressing fate: during the night they were flung out into space from the top of the terrace so that they fell into the Golden Horn. This was referred to as the 'kourban' of the 'papoutchs', the sacrifice of the 'babouches'.

On perfectly clear, cold nights it was a delight, climbing up the old wooden stairs that creaked beneath our tread as it lead us to up to the roof; there, by the the beautiful light of the moon ('mahitabda'), once we were sure that everyone in the neighbourhood was asleep, we would carry out the 'kourban' and send the condemned 'babouches', one by one, spinning through the air.

Will the 'papoutch' fall into the water, or onto the mud, or even on the head of a prowling cat?

The sound we heard in the deep silence would indicate which of the two of us had guessed rightly and had won the bet.

It was good being up there, all by ourselves, so far from humanity; we felt easy in our minds, and we'd often walk about on the carpet of white snow and gaze down upon old Stamboul as it slept. It was out of the question for the two of us to enjoy a walk together in broad daylight, as so many others do, arm in arm in the sunshine, taking it all for granted. The roof was where we took our walks; there we could breathe the sharp, pure air of beautiful winter nights, accompanied by our discreet friend, the moon, who sometimes sinks slowly over the infidel countries of the west, or at other times rises in the east, all red, outlining the distant silhouettes of Scutari or Pera.

XXXVII

'Is this the end, Lord, or the beginning?'

(VICTOR HUGO, 'Songs of Twilight')

Great activity on the Bosphorus. Transport vehicles arriving and departing, full of soldiers headed for the front. They come from everywhere, soldiers and reservists alike, from deep into Asia, from the frontiers of Persia, even from Arabia and Egypt. They are hurriedly kitted out and sent off to the Danube or to camps in Georgia. Resounding fanfares and awesome shouts in honour of Allah announce their departure. Never has Turkey seen this many men, so brave and so resolute: these men under arms. Allah alone knows what will become of these multitudes.

XXXVIII

Eyoub, 29 January 1877.

I would never have forgiven their Excellencies if their diplomatic lampooning had disturbed my private life.

Happily, I find myself back in my small, secluded house, which, for a brief time, I feared I'd be leaving.

It's midnight; the moon sends its blue light to drift across my page, and the cocks have begun their nightly song. Here in Eyoub I am so far from my own kind, and at night how solitary it is; but by the same token, how peaceful. I often find it hard to believe that Arif-Effendi

is me; but I'm so weary of the me that I've known for twenty-seven years that I rather like being able to assume a little of this other.

Aziyade is over in Asia with the rest of her harem, visiting another harem in Ismidt; she'll be returning to me in five days time.

Samuel is down on the floor beside me, sleeping like a babe. Earlier in the day he saw a drowned man who'd been fished out of the water. Apparently the body was a ghastly sight and frightened Samuel so much that, to be on the safe side, he transferred his mattress and blanket into my room.

Tomorrow from first light the reservists who are off to the front will be making a din and the mosques will be crowded. I would willingly go with them and also be killed somewhere or other in the service of the Sultan. It's a fine, stirring thing, this struggle by a people that doesn't wish to die; and I feel for Turkey a little of that fervour that I would feel for my own country were it to be similarly threatened and put in mortal danger.

XXXIX

We were sitting, Achmet and I, in the square outside Sultan Selim's mosque. We were letting our eyes follow the old stone arabesques that climbed, twisting and turning, up the grey minarets, and letting them follow too the smoke from our 'chibouks' as it rose, spiralling in the pure air.

Sultan Selim's Square is surrounded by an ancient wall in which are set at intervals doors that curve to a point. It's rare for anyone to come this way; there are a few tombs, sheltered by cypress trees; it's a fine part of the city, and it could be Turkey as it was two centuries ago.

'Me,' said Achmet, wearing a rebellious face, 'I know

very well, Loti, what I'll do when you're gone: I'll have myself a merry old time and get drunk every day. I'll get an organ grinder to follow me around, providing me with music from morning till night. I'll run through all my money--but I don't care ('zarar yok'). I'm the same as Aziyade, when you've gone it will be the end of your friend Achmet too.

So I had to make him swear to be sensible--no easy matter.

'Will you swear an oath for me in return, Loti? Will you swear that

when you're married and rich, you'll come and find me and let me be your servant over there? You're not to pay me any more than you pay me in Stamboul, but I will be near you, and that's all I ask.

I promised Achmet that I would give him a place under my roof and that I would entrust my children to his care.

The prospect of bringing up my little ones and topping them off with a fez was all it took to bring him round. We frittered away the whole evening on educational schemes based on extremely original methods.

XL

PLUMKETT to LOTI

'My dear friend,

'The reason for my not writing to you is simply that I had nothing to tell you. When that's the case, I tend to keep quiet.

'And really what was there I could talk about? Would you want to know that I had a lot of things on my mind, none of them pleasant; that Lady Reality had me in her clutches, an embrace mighty difficult to shake off; that I was languishing somewhat dolefully in the company of naval and colonial gentlemen; that the bonds of mysterious affinities, which at certain times have brought me such an intimate knowledge of all that is lovely and kind, were broken?

'I'm sure you'll understand all this very well, because I've seen you more than once plunged in the same state.

'Your nature is very much like mine, and that neatly explains, I think, the great sympathy I felt for you almost from the first.--An axiom: 'What we like best in other people is ourselves.' When I meet another me, my own powers redouble; it would seem that the like forces in each person combine and that sympathy is no more than the desire, the impulse towards this increase of forces which for me is synonymous with happiness. With your kind approval I'll call it: 'the great sympathetic paradox.'

'The language I'm using here could hardly be called literary. I'm well aware of it: I employ a vocabulary borrowed from dynamics, hence it's altogether different from that used by our best writers; however, it's very good for getting my ideas across.

'We encounter these sympathies in a host of different ways. Need I ask a musician, such as yourself, what it is that prompts you to feel them? What is a sound? Quite simply its a sensation which arises in us as a result of a vibration being transmitted by the medium of air to our tympanum and acoustic nerve. What happens in our brain? Try to picture this peculiar phenomenon: you register a run of certain sounds, you hear a melodic phrase that you find pleasing. Why is this so? Its because the succession of musical intervals that go to make up the phrase—otherwise known as the ratios of the numbers of vibrations

of the resonant body--are marked in one set of figures rather than another; change these figures, and your sympathy is no longer aroused; you, yourself, will say that the phrase is no longer a musical phrase but merely a succession of incoherent sounds. When several sounds are played simultaneously, you receive an impression which will be happy or sad: it all comes down to the matching of figures, which in turn constitutes a sympathetic match between an exterior phenomenon and your sensitive self.

'There are real affinities, the ones between you and certain successions of sound, between you and certain brilliant colours, between you and certain shimmerings of light, between you and certain lines, certain shapes. Although these relationships between all these different things and yourself have been confirmed, they are too complicated, as in the case of music, to be put into words. All the same, you can tell they exist.

'Why is it we fall in love with a particular woman? Very often it's simply a matter of the curve of her nose, the arc of her eyebrows, her oval face--what would I know? But whatever it be, it corresponds with another unfathomable that is in us and plays the devil with our imagination. Why the surprise? Half the time our love is due to nothing more than that.

'You tell me this woman of yours possesses a nobility of character, wherein lies the true reason for your love ... Alas! beware of confusing what is in her with what is in you. All our illusions stem from this: crediting others with attributes we find pleasing yet exist nowhere but in ourselves. Building a shrine to the woman we love and considering one's friend to be a man of genius, have a common origin.

'I've been in love with the Venus de Milo and with one of Correggio's nymphs. It certainly wasn't the charm of their conversation and thirst for intellectual exchange that attracted me to them; no, it was the physical affinity, the only love known to the ancients, the love that produced artists. These days everything has become so complicated ne hardly knows which way to turn. Nine-tenths of the population no longer understand anything at all.

'That said, let's move on, Loti, to your description of yourself. There is an affinity between the way things are and the way you are. By disposition you have an enormous appetite for artistic and intellectual pleasures, and you're unable to be happy unless you're surrounded everything that can satisfy your sympathetic needs, which are considerable. There is no happiness for you beyond these emotions. Take away the means by which you obtain them and all you'll be left with will be the reality of being a hapless exile.

'A peson who can handle emotions of such a high order, as opposed to the great mass of us who can't, will have scant regard for anything less. What attraction can there be, then, in a good dinner, a day's shooting, a pretty girl, for anyone who, when reading the poets, has shed tears of rapture, who in delight has surrendered to the flow of a sweet melody, who has immersed himself in that deep reverie, that isn't thought, and goes beyond feeling, and cannot be expressed in words.

'What pleasure, then, in eyeing ordinary-looking faces as they go by, common faces portraying every shade of stupidity; what pleasure in ill-proportioned bodies imprisoned in culottes or in black clothes, the whole lot of them swarming about the muddy pavements , around filthy walls, window-boxes and shops.

'It narrows one's imagination and one's ability to think.

'What prompts those round you to talk with you if its not the existence of a harmony between your thoughts and the ones they have and are trying to express?

'If you can let your mind project itself into space and time; if it can encompass the infinity of simultaneous events which are happening over the whole surface of the earth, which is no more than a planet revolving round the sun,--itself only one of an untold number out in space; if you can appreciate that this simultaneous infinity is but an instant in eternity, which is yet another infinity, and that all this would appear different to you depending on what observation point you chose, there being an infinite number to choose from; if you can reflect on the reason for it all, the fundamentals of all these things are beyond your knowing; and if you tax yourself with these eternal problems: what does it all mean? what am I,in the midst of infinity?

'There's every chance that there'll be no common ground, intellectually, between you and those around you.

'Their conversation will have no more effect on you than if you were listening to a spider telling you how some nasty brute of a feather duster had destroyed part of its web; or, perhaps, a toad, informing you of the fact that he'd just inherited a great heap of rubble which would provide him with a very comfortable home.

'(A gentleman told today that he's had a run of bad harvests, and that he'd been left a big house in the country.)

'You've been in love--it could be you're still in love; at any rate you've experienced the feeling that there exists a kind of life that's altogether special, part of your being that is private and enables you to see everything in a totally new light.

'A kind of epiphany appears to be taking place; you might liken it to being born a second time, because from then onwards you find you're getting more out of life, and are a whole person. All the ideas and feelings that have lain dormant awake and brighten like the flame under the punchbowl when you shake the rim. (Literature of the future!)

'In short, you come into your own, you're happy, and all that went on prior to this new found happiness is engulfed in a kind of night. You feel you're in limbo; in terms of your everyday life you're more a young boy than a young man. When you're in love, you can describe the feelings you go through only at the very moment you experience them--and I have to say, I'm not having those sorts of feelings right now.

And yet, what have we here? by Jove! I'm getting

quite carried away, tossing all these ideas about; I'm getting excited, I'm going out of my mind! ... How good it is to love and be loved, to know that one of the most beautiful souls has understood yours; that there's someone who will recount her every thought and action to you, so that you are, in fact, her centre, her destination, and this is due to the fact that the ways in which she lives, thinks and acts are every bit as subtle and intricate as your own. Now that's what makes us strong; that's how men of genius are made.

'And then we have the graceful image of woman whom we love, which is, perhaps, not so much a reality as the most flawless picture your imagination can produce; and there's that blend of impressions, physical and moral, sensual and spiritual, not to mention those impressions which completely defy description and can only be called to mind by those who are already acquainted with them—impressions, which result from a mysterious association of ideas such as the least little thing that belonged to your beloved, or hearing her name mentioned, or simply when you see it written on a sheet of paper, or it could be one of a thousand other sublimely absurd things that may well turn out to be be the one that you value more than anything else in the world.

'Then comes friendship, a stricter kind of sentiment, one that's on a more solid footing, basing itself on everything that's highest in us, that part of us that's purely intellectual. Such a privilege it is being able to speak candidly to someone who understands you, through and through, not just part way, someone who can finish off what you're aiming to say with the very word that was on your lips; someone whose reply can bring forth in you a torrent of images, a flood of ideas. A hint from your friend can tell you far more than a string of sentences, the reason being that you're used to his way of thinking. You understand every emotion that drives him, and he's aware of it. You are two intelligences that add to each other and complement each other

'One can be sure that anyone who has experienced the very things I've just referred to, and who experiences them no more, is greatly to be pitied.

'Here, then, am I, bereft of affections, no one to spare me a thought. What good are ideas if there's no one to tell them to? What good is talent if there's not one person in the world whose esteem I value above the rest. What good is wit among people who won't be able to understand me?

'So I just drift along; I've had disappointments ; in fact, fresh ones arrive every day. It's clear to me now that nothing in this world endures, and that there's nothing one can count on, nothing at all.

My nerves lack tension, the level of my thoughts is low, the me is diminishing to such an extent that, when I'm alone I sometimes wonder if I'm awake or sleeping. My imagination ceases. No more castles in Spain. The same applies to hope. You lapse into bravado, you talk in an offhand way of the many things, that, if they wren't so laughable, they'd make you weep.

'There's not a thing you love, yet you were born to love everything: there's not a thing you believe in, though one might still believe in everything; time was when you were fit for anything, now one is fit for nothing.

'To have within me an exuberance of faculties and yet feel I'm letting it all slip away; an excrescence of sensibility, a superfluity of feelings that I don't know what to do with: its dreadful! Life under those terms is a daily agony: an agony that certain pleasures can alleviate for a short while (your circus girl, the odalisk Aziyade, and the other Turkish floosies); but it only means I lapse back again more bruised than ever.

'There you are, there's your declaration of faith analysed, developed and considerably enlarged by the odd character sitting here writing to you.

'To conclude this lengthy stretch of gibberish, just let me say this: I take a very keen interest in you, less, perhaps, because of what you are, than because of what I think you might become.

'What made you choose muscle building to distract you from your sorrow when it will kill in you the only thing that can save you? You are clown, acrobat and fine marksman; it would have been far better, my dear Loti, if you'd been a great artist.

'In addition to all I've said, I'd like to plant in your mind this idea

in which I do have faith: there is no anguish that does not have its remedy. It lies with our reason to find that remedy and apply it, in a way that is appropriate to the nature of the ill and the temperament of the subject.

'Despair is a totally abnormal state to be in; its a sickness that's just as curable as umpteen others; its natural remedy is time. No matter how unhappy you are, see to it that there's always a little corner of you that you'll not allow the illness to invade; this little corner will be your medicine box.--Amen!

'PLUMKETT.

'Tell me about Stamboul, the Bosphorus, 3-tail pashas, etc.I kiss the hands of your odalisks, and remain affectionately yours.

PLUMKETT.'

XLI

LOTI to PLUMKETT

'Had I told you, dear friend, that I was unhappy? I don't think so, and if I had, then I was certainly mistaken. And it couldn't be further from the truth. I came home this evening telling myself that I was one of the lucky ones of this world, and that the world itself was a very fine place. I returned on horseback on a beautiful January afternoon; the setting sun was gilding the dark cypresses, the ancient, crenellated walls of Stamboul and the roof of my secret house where Aziyade was waiting for me.

'A brazier was warming up my room that was already heavily perfumed with essence of roses. I bolted my door and sat down with my ankles crossed--a pleasant way to sit, something you have yet to experience. My servant, Achmet, prepared two hookahs, one for me, the other for himself, and then placed at my feet a copper tray containing a burning seraglio pastille.

'Aziyade, in a low voice and tapping on her tambourine, started to sing the song of the jinees; the smoke began to form bluish spirals in the air and gradually I lost all consciousness of life, of the sorry lot of humanity, while I gazed on the three, kind, friendly faces near me: my mistress, my servant, and my cat.

'What's more, we get no intruders, no unexpected or unpleasant visitors. A few Turks may pay me a discreet visit when I invite them, but my European friends haven't the faintest notion of where I live, and the ash-wood trellising guards my windows so closely that there's not a single moment in the day when any nosy person could peep within.

'It takes an oriental, my dear friend, to tell you what 'being at home' means. In your European dwellings, that are open to all and sundry who pay you a visit, you can be no more 'at home' than you could, standing out in the street; you lay yourself open to the over-inquisitiveness of troublesome friends as well as to downright snoopers; the charm and mystery of domestic inviolabilty as understood over here is a closed book to you.

'I'm happy, Plunkett; I take back all the moans I was fool enough to send you ... and yet I can still feel the pain from all that

lay broken inside my heart: I feel that the present is only a respite in my destiny, that there's something ghastly hovering over my future, that today's good fortune will inevitably usher in a dreadful tomorrow.

Even here, when she is close to me, I have these moments of appalling sadness comparable to the unfathomable feelings of distress that often gripped me when I was a child and night was falling.

'I'm happy, Plunkett, and I'm even feeling younger; I'm no longer that overgrown lad of twenty-seven, who had knocked about so much, seen so much of life and had done every stupid thing possible in every country imaginable.

'It would be difficult to decide which of the three is most like a child: Achmet or Aziyade, or even Samuel. I might be old and sceptical, but when I was with them I looked like one of those Buldwer characters who lived ten human lives without it showing on their face: in my case an old, wearied soul lodged within the body of a twenty year old.

'But their youth refreshed my heart, and you are right, perhaps I could, even now, believe in all things, I who thought I no longer believed in anything ...'

XLII

It was one January afternoon; the entire sky over Constantinople looked grim; a cold wind hurried along a winter drizzle, and the daylight was as feeble as they usually are in England.

I was riding down a long, white road; on either side of me endless walls of polished stone rose sheer some thirty feet and looked as inaccessible as the walls of a prison.

At one point along the road an arched bridge, built of grey marble, rose into the air; it was supported by curiously carved marble columns and served as a thoroughfare between the right part and the left part of these depressing constructions.

These were the walls of the Tcher. Seraglio. On one side were the gardens, on the other the palace and the pavilions. The marble bridge

allowed the beautiful sultanas to pass to and fro between the two and to avoid being seen from the outside.

Opening onto the palace ramparts and set well apart from one another were no more than three gates, each framed in grey marble, and closing by means of iron tongues, all chased and gilded.

Actually, it was the high majestic gates that set me wondering what treasure might lie behind those monotonous walls.

Soldiers and black eunuchs guarded these forbidden entrances. Just the look of the porticoes seemed to warn you that to it was dangerous to try to pass through. The marble columns and friezes interlaced in the Arab style were covered in strange patterns and mysterious scrolls.

There was a mosque of white marble with gilded dome and crescents that backed onto some dark coloured rocks and scrub. It was as if some peri had waived its wand and, without disturbing the rusticity of the surroundings, had caused the mosque to suddenly appear in all its snowy whiteness.

A rich carriage drove past. It contained three Turkish ladies, none of whom I knew. One of them displayed, beneath her transparent veil, what seemed a rare kind of beauty.

Having two eunuchs riding behind them as escorts showed that the ladies were of notable rank.

Behaving atrociously in public is the way with all upper class 'hanums'. There's hardly anything that will deter them from sending the most enticing or sneering looks to the European men they pass in the street.

The especially pretty one gave me such an indulgent smile that I pulled

my horse round to follow her.

Then began a two-hour-long ride, during which the lovely lady exhibited for my benefit, through the open window of her door, the whole range of her delightful smiles. The carriage fair sped along and I escorted it all the way, sometimes in front, sometimes behind, slackening speed, or galloping to increase it.

The eunuchs--they're often tremendous in comic operas--cast benevolent eyes over what was going on, and without varying their distance from the carriage, continued to trot along, completely unperturbed.

We went through Dolma-Bagtche, Sali-Bazar, Top-Hane, Galata's rumbustious quarter,--and then over the Stamboul bridge, and on through gloomy Phanar and dark Balat. Finally we came to Eyoub and turned into one of the old streets. Outside an old conak (inn) that looked opulent and dark, the carriage stopped and the three ladies alighted.

Before entering her residence, the beautiful Seniha (I learnt her name the following day), turned round to give me a parting smile. She was much taken by my boldness, and Achmet predicted that my adventure would end badly.

XLIII

Turkish wives, especially the leading ones, set very little store by the faithfulness they owe their husbands. The fierce surveillance carried out by certain types of men, and the terror of being punished, are indispensable for keeping them in check. Constantly idle, bored out of their minds, physically dominated by the solitude of the harems, wives are capable of giving themselves to the firstcomer—the servant who happens to come along, or the boatman who takes them on the water,

provided, of course, that the man is good-looking and they like him. The wives are all tremendously curious about young people from Europe, and there are those among the latter who might sometimes do themselves

a bit of good if they wanted to, if they dared or, better still, if they found themselves in circumstances where they could put it to the test. My position in Stamboul, my knowledge of the language and customs, my door that's out of sight and swings noiselessly on its old hinges— these were the things that boded well for those kinds of enterprises, and, had I wanted it, my house could easily have become the meeting-place for all the lovely, harem wives who were at a loose end.

XLIV

A few day later, a great thundercloud crashed down on my peaceful house, a most frightful thundercloud that worked its way between me and the woman that I had, in fact, never ceased loving tenderly. Aziyade was outraged when I told her of a cynical plan I had. Without a tear in her eyes or tremor in her voice she resisted me with a strength of will that aimed to master mine.

I'd announced to her that I didn't want her to come the following day, and that another woman would be taking her place for a few days; after that, she should return to me and go on loving me as she'd done before, despite any blow she felt to her dignity, which she should cancel from her memory.

She knew this Seniha, who was renowned in the

harems for her scandals and for not getting caught. Aziyade hated this creature that Behidje-hanum heaped curses upon. The idea of being brushed aside for the sake of this woman filled her with bitterness and shame.

'It's my last word, Loti', she said. 'If that Seniha woman comes here, it will be over between us, and I shan't even love you any more. My soul is yours and I will belong to you; you are free to do as you please. But, Loti, this will be the end; I'll likely die of grief, but I will never see you again'.

XLV

An hour later, when love had prevailed, she'd consented to my madcap compromise. She left, giving me her word that she would return--later,

when the other one had gone and it would please me to send for her.

Aziyade departed, her cheeks flushed, her eyes dry, and Achmet, falling in behind her, turned round and informed me that he was leaving for good. The Arab curtaining across the door of my room fell back behind them and I could hear their 'babouches' trailing across the mats as far as the stairs. There they paused. Aziyade had collapsed on the steps and had burst into tears. The sounds of her sobbing travelled back to me in the silence of the night.

Nevertheless, I kept to my room and let her go.

I'd just been telling her, and it was true: I worshipped

her, and her alone, and I had no love whatsoever for this Seniha; with me it was simply a fever of the senses, and it was this that was drawing me towards this wildly intoxicating nobody. It distressed me to think that once Aziyade was settled back behind the walls of the harem, she may not, in fact, wish to see me ever again, in which case she would be lost to me forever, and no power on earth could restore her to me. When I heard the house door close behind them, I felt a pang that defies description. But in spite of that, the thought of the creature who would soon be with me, set my blood aflame. So I held my ground and didn't call them back.

XLVI

The following evening my house was decked out and perfumed to receive this lady of quality, who had expressed a genuine desire to see inside my solitary abode. The beautiful Seniha arrived very mysteriously, on the stroke of eight, an unearthly hour for Stamboul.

She removed her veil and the common grey woollen cape, which she had chosen as a precaution, and let fall the train of a French gown that failed to captivate me.

It was in doubtful taste, more expensive than fashionable, and didn't suit her. She read my thoughts. Passing off her failure to impress me, she seated herself gracefully and let the words flow from her. Her voice lacked charm and her eyes roamed inquisitively all over my room, which she applauded for its originality and its look of comfort. She kept commenting on the oddness of the life I led and didn't hold back from asking me lots of questions that I avoided answering.

All the while, I studied Seniha-hanum ...

She was a most splendid creature: her youthful skin smooth and velvety, her parted lips red and moist; she carried her head thrown back, haughty and proud, fully aware of her sovereign beauty.

An overwhelmingly fierce sensuality showed in her smile, in the slow turning of her dark eyes, half-hidden beneath the fringe of her eyelashes. Seldom had I experienced having such a beautiful woman there beside me, awaiting my pleasure in the warm solitude of a perfumed room; and yet, inside me an unexpected conflict was gaining ground. My senses were battling against that less definable something generally known as the soul, and the soul was battling against the senses. It came home to me right there and then that I was truly in love with my darling, whom I had driven away; my heart was overflowing with tenderness and remorse. The beautiful creature sitting beside me inspired me more with disgust than with love. I had desired her; she had come; it was up to me now to have her; I had asked for no more than that; now her presence was abhorrent to me.

Conversation grew feeble, causing Seniha to make tongue-in-cheek remarks. I steeled myself against my own inclinations, utterly determined that this woman should not succeed in deflecting me.

'Madam,' I said, continuing to speak in Turkish, 'when the times comes when it will pain me to see you leave (and I hope it will not be yet for some considerable time) will you allow me to escort you back to your house?

'Thank you, but no,' she replied. 'I have someone.'

This was a woman who took no chances: a kind-looking eunuch, doubtess accustomed to his mistress's escapades and prepared for all eventualities, was standing outside near the door to my house.

As she was going out, the great lady gave a jeering laugh which brought a flush of anger to my face and I was close to grabbing her lovely round arm and bringing her back.

But I resisted, telling myself that I hadn't gone to any trouble on her account and that of the two roles we had played the more bizarre one definitely hadn't been mine.

XLVII

Achmet, the same Achmet who was never going to return, showed up the next morning at eight.

He'd adopted a very surly expression and greeted me coldly.

My desription of Seniha-hanum's visit soon had him in stitches; he came to the conclusion, as I knew he would, that it was all of a piece with my usual behaviour. 'You're a one and no mistake ('tchok chéytan'),' he

said and sat himself down in a corner to savour the account and to laugh some more.

Afterwards, whenever we were riding together and encountered Seniha-hanum's carriage, he would give her such taunting looks that I was obliged to tackle him about it and give him stern lectures.

XLVIII

I sent Achmet off to Oun-Capan to find Kadidja and give the monkey-like

confidante details of the night Seniha was my guest. Kadidja was to tell Aziyade that I begged her forgiveness, and that my dearest wish was for her to come me that very evening.

At the same time I sent three children into the countryside to get me fresh green branches and sprays, and basketfuls of narcissi and jonquils. On the day of her return I wanted the old house, for once, to look joyful and festive.

That evening when Aziyade came in she was met by a carpet of flowers

that stretched from the threshold of the house to the door of our chamber; a thick layer of fragrant jonquil flowers heads, plucked from their stalks, covered the floor; their sweet perfume was intoxicating, and the steps where she had shed tears were no longer visible.

Neither comment nor reproach came from her rosy lips, she simply smiled as she looked at all the flowers; she was intelligent enough and more to see at a glance what words of mine they were silently conveying; and her eyes, dark-ringed from weeping, glowed with intense joy. She stepped calmly and proudly onto the flowers like a little queen resuming possession of her realm that was thought to be lost; or like apsaras moving through the floral paradise of Hindu deities. But be they asparas or houris, there could be none as lovely, fresh, graceful, delightful...

The Seniha-hanum episode was ended; its only result was to make us love eachother more deeply.

XLIX

It was winter, the time of evening prayer. The muezzin chanted the eternal song, and we were tucked away, just the two of us, in our secret Eyoub abode.

I can still see her, darling little Aziyade, sitting on the floor on a red and blue carpet--since taken from us by the Jews--bolt upright, cross-legged in her Asian silk pantaloons, and looking very serious. She wore that particular expression that verged on the prophetic and contrasted so markedly with the sheer youthfulness of her features and the artlessness of her thinking; it was the expression she adopted when she wanted to drum into my head some deduction of her own that more often than not derived from some oriental parable and was, in her mind, utterly compelling.

'Bak, Lotim,' she said, fixing her deep eyes on me, 'Katebtane parmak bourada var?'

And she held up her hand with the fingers outspread.

(Look, Loti. Tell me, how many fingers are there here?)

I laughingly replied:

'Five, Aziyade.'

'Yes, Loti, only five. And yet none are the same. 'Bou, boundan bir partcha kutchuk.' (This one, the thumb is a little shorter than the next; the second a little shorter than the third, etc.; whereas this one, the last one, is the shortest of all).

It was indeed very tiny, Aziyade's shortest finger. Her nail, apart from the deep pink at the base from which it grew, was like all the others, stained with henna a beautiful orange red.

'Well,' said she, 'in just the same way, but all the more so, none of Allah's creatures—and they are much more numerous—are alike; every woman is different, every man is different ...

The aim of this parable was to prove to me that even though women I had loved in the past had found it possible to forget me, and even though friends had deceived me and deserted me, it was wrong to judge every man and every woman by their example. She, Aziyade, was not like the other women; she would never forget me; Achmet likewise would never cease to love me.

'So stay with us, Loti, do stay with us ...'

Then her thoughts turned to the future, that dark unknown which held a fascination for her.

Old age, that thing far, far away in the distance, that she could barely picture... But why not grow old together and go on loving eachother-- loving eternally in this life and beyond?

'Sen kodja,' she said (you will get old); 'ben kodja, (I will get old) ...'

Characteristically this last sentence was, not so much spoken as mimed.

The words 'I will get old' were conveyed by her forcing her young voice to sound cracked, and for several seconds she sat huddled up like a little old woman, bending that body of hers that was all bursting with the glowing freshness of youth.

'Zarar yok' (it's of no importance), was her conclusion. 'Loti, we will always love eachother.'

L

Eyoub, February 1877.

Odd, when I think about it, how our story began!

All the reckless actions, and blunders we crammed into our daily lives for a whole month, with the purpose of achieving a result which was intrinsically impossible.

Getting dressed up in Salonika as a Turk and wearing clothes that offended any reasonably observant eye by the very exactness of its details; going about the city like that when a simple question from a passer-by could have exposed and endangered the impudent infidel; to court a Moslem woman beneath her balcony: an enterprise without parallel in the annals of Turkey. And all of that, good God, more to relieve the boredom of living, and to set oneself a little apart and out of the ordinary in the eyes of one's idle comrades, more to hurl a challenge at existence, more a piece of bravado than a gesture of love.

Then success comes along and crowns this surfeit of

recklessness that has been brought about by means that should really have resulted in tragedy.

Perhaps it goes to show that it's only those things reputed to be insane that end well, and that crazy people have their own source of good fortune: a God that looks after the foolhardy.

... In her case it was curiosity and apprehension that first aroused feelings in her heart. In the beginning it was curiosity that had kept those big eyes of hers gazing, more out of astonishment than love, through the balcony trellises.

Initially he'd made her shiver, this stranger who could alter the way he dressed just as Proteus of old could change form, and stand there beneath her window, all done up in gilded Albanian attire.

Later the thought had come to her that he must love her very much, her, the bought slave-girl, the lowly Aziyade, because he so recklessly risked his life simply to gaze at her. Poor child, she had no reason to suspect that this young man with such a boyish face had already enjoyed to excess all that life had offered him and had come in search of some novel experience, bringing with him nothing but a world-weary heart; she'd told herself that only good could come of being loved like this,--and very gradually she'd slid down the slope that inevitably led her into the arms of the infidel.

There had been no one to teach her a single moral principle that might have put her on guard against herself,--and little by little she'd yielded to the spell of the only poem of love that had been sung for her, and to the tremendous attraction of its danger. In the beginning she had proffered her hand through the window bars of the 'yali' by the Monastir Road; and then her arm, and then her lips, until one evening when she had opened wide her window and had descended into her garden ,--like Marguerite, a picture of youth and spontaneity.

Her soul, like Marguerite's, was pure and virgin, even though her child's body, an old man's purchase, was no longer so.

LI

These days we proceed more warily and take no risks; we have a thorough knowledge of Turkish customs; we know all the streets to avoid in Stamboul; we have perfected the art of dissimulation. Yet we still tremble whenever we meet somewhere, and the memories of those first months in Salonika seem memories of dreams.

Often, as we sit together before the fire, like two children who have learned how to behave and now speak of their past misdemeanours in hushed tones, we talk of the restless period we spent in Salonika, of the hot thundery nights when we roamed the countryside like criminals, --or drifted about on the sea as if we were insane--all the while incapable of exchanging a single thought, let alone a single word.

What is still most unusual about it all is the fact that I love her.--The 'little blue flower of simple love' has bloomed anew in my heart, and I attribute it solely to her burning young passion. To the very depths of my soul I love her and I worship her ...

LII

One fine Sunday in January, I was nearing home in the cheerful winter sunshine when I saw about five hundred people hurrying to and fro with water pumps.

'What's on fire?' I asked urgently.

I'd always had a presentiment that my house would catch fire.

'Go quickly, Arif!' an old Turk replied 'Go quickly, Arif! It's your house!'

I had to deal with a type of emotion I'd never met before.

I displayed no feeling at all as I approached the little abode that we had furnished for eachother, she for me, and I for her, and with so much love.

The hostile and menacing crowd parted to let me through; old women, already worked up into a frenzy, flung insults at me and stirred up the men; people had smelled sulphur fumes and seen green flames; I was accused of sorcery and casting evil spells. Their original suspicions had only lain dormant, and I was suffering the consequences of being an implausible and disquieting figure; I had no one to turn to and was without any resources.

I walked slowly up to the house. The doors had been pushed in and the windows broken; smoke was coming out through the roof; looters had been busy, one of those ghastly mobs that crop up in Constantinople whenever there's the chance of a fight. I stepped into my home and was greeted by a rain of dirty water and soot, flakes of charred plaster and floor boarding that was still alight ...

However, the fire had been extinguished. One room burnt out, one floor, two doors, and a partition. Aided by a big dose of sang-froid I had got the situation under control. The 'bachibozouks' had forced the pillagers to surrender their booty, then they'd cleared the square and dispersed the crowd.

Two armed 'zapties' stood guard at my forced-in door. I entrusted to them the care of my belongings and took the ferry across to Galata. I was going to fetch Achmet, a lad wise beyond his years, whose presence would be invaluable to me amid all this confusion.

An hour later I arrived in the centre of rowdy conglomeration of bar-cafes; I wasted time checking at 'Our Madame's' and in all the dives: but that evening Achmet was nowhere to be found.

So all I could do was go back home and and sleep alone in my windowless, doorless room, balled up, on account of the deadly cold, in wet blankets that smelled of burning. I got little sleep, and my thoughts were gloomy; it was one of the most unpleasant nights I'd ever spent in my life.

LIII

The next morning Achmet and I assessed the damage; it could have been a lot worse, and there was nothing that couldn't be repaired. The room that had been gutted had been unfurnished and unoccupied; for what it had accomplished, I thought the fire might simply have been staged; the most trifling items kept popping up everywhere, in disorder and dirty but still intact.

Achmet was a bundle of energy; he had three old Jewesses at his command doing the scrubbing and tidying; and some of the goings on were absolutely hilarious.

During the following day the whole house was swept, cleared of debris, washed, dried and made nice and clean. Where two rooms had been there was nothing save a black gaping hole; but that detail apart, the house had regained its equilibrium, and my room its original elegance.

That very same evening my rooms were laid out for a grand reception;

lots of trays laden with narguilhés, ratlokum, and coffee; even a two-musician orchestra: a drummer and an oboeist.

It was Achmet who had urged all the expenditure and had arranged everything: at seven o'clock I was to receive the authorities and the notable citizens who would be deciding my lot.

I was fearful of having to reveal my identity which would mean my claiming the protection of the British Embassy: I was in a high state of bewilderment as I awaited my company.

For my adventure to be ended in this way would have inevitably resulted in an order from a higher level, cutting short my stay in Stamboul, and this solution I dreaded even more than Ottoman justice.

I can still see them all, the entire scene: some fifteen or twenty people seated on my rugs; landlord, notables, neighbours, judges, police, and Dervishes; the racket from the orchestra; and Achmet issuing cups of coffee and mastic filled to the brim.

It all hinged on whether I could clear myself of any charge of arson or witchcraft; and whether I went to prison or paid a heavy fine for having nearly burned Eyoub to the ground; lastly there was the question of my landlord's indemnity and how much the repairs would cost me.

In Turkey, it's best, for the most part, to rely on one's own initiative; and, generally speaking, daring will pay dividends and self-confidence can lead to success. All evening I maintained a dignified front, and brazened things out; Achmet kept everyone topped up and deliberately got everyone's questions tangled up with their reasons, and carried it off magnificently;--the orchestra raged on and after a couple of hours the situation was at its peak: my guests were arguing with one another totally at cross-purposes; my case was entirely forgotten.

'There you are, Loti,' said Achmet, 'they're right on time and its all my doing. In the whole of Stamboul you'd never find another like your old Achmet; truly, I'm very valuable to you.'

The whole business was both complicated and comic,--and Achmet's madcap

mirth was contagious; I yielded to a need that would not be ignored and, without further ado, I dived onto the palms of my hands and performed a couple of clowning tricks in front of an astounded audience.

A delighted Achmet hit upon a similar diversionary tactic; bowing extravagantly, he handed each visitor his pattens, cloak and lantern, and the meeting broke up without anything being decided.

End and Moral: I didn't go to prison; and I didn't pay a fine. My landlord saw to the cost of repairing the house, giving thanks to Allah for having spared the other half of it; and I remained the spoilt child of the neighbourhood.

It's hard to imagine a fire starting up, all by itself, in the middle of a locked house, and the initial cause of it continues to be a mystery.

LIV

'What Nature offers on these shores is forgetting...

Whoever plunged into the Ocean of the heart

Found Repose in such Annihilation,

For nothing else abides there...

(FERIDEDDIN ATTAR, Persian poet.)

Izeddin-Ali-effendi was holding a reception at his house in old Stamboul: smells of perfume and 'tembaki' tobacco, sounds of tambourines and men's voices dreamily chanting weird oriental melodies.

These evening parties, which had struck me at first as being weird and barbaric, have gradually grown on me and I've given similar ones at my own house that offered the heady mixture of tambourines, perfumes, and swirling tobacco smoke.

Guests arrive at Izeddin-Ali-effendi's receptions in the evening and don't leave until its already broad daylight. On a snowy night in Stamboul distances have to be considered, and Izeddin's concept of hospitality is a very wide one.

Izeddin-Ali's house, old and shabby on the outside, contains within its dark walls hidden glories of oriental luxury. Izeddin-Ali, be it said, is one of that exclusive group that venerates everything 'eski', that is: everything that recalls a past long gone and bears its seal.

I knock at the heavy ironbound door, which is opened soundlessly by two little Circassian slave girls.

I extinguish my lantern and remove my shoes. These are courtesies required by custom in Turkish homes, where mud from the street must never be brought indoors. The precious carpets that are handed down from father to son feel the tread only of 'babouches' or bare feet.

The two little slave girls are about eight years old; they can be sold, and this they know. Their bright smiles and even features are enchanting; their baby locks are threaded with flowers and piled high on their heads. Respectfully they each take my hand and gently put it their forehead.

Aziyade, who was once a little Circassian slave girl herself, has retained the same manner with me of of expressing submission and love ...

I ascend the dark stairs that are covered with sumptuous Persian carpets; quietly the mother-of-pearl ecrusted door to the 'haremlike' begins to open and when it's half way I see I'm being watched by female eyes.

In a large hall, where the carpets are so thick I could be walking on the back of a Kashmir sheep, five or six young men are seated cross-legged in postures of smiling indifference and peaceful reverie. A large chased-copper vessel, full of glowing embers imparts to the air around a gentle if somewhat heavy warmth that induces sleep. From the carved oak ceiling, heavy clusters of candles are enclosed in opalescent tulip shaped globes that transmit only pink light, soft and dim.

Chairs, like women, are absent from Turkish parties. Nothing except very low divans, covered in rich Asian silks; cushions brocaded in satin and gold; silver trays on which lie long pipe stems of jasmine; and then, some little eight-panel stands for the hookahs that are finished off with big balls of amber encrusted with gold.

Not everyone is admitted into Izeddin-Ali's house, and those who are have been carefully chosen; not from the pashas' sons, who loaf about the Paris boulevards, witless and flashy; no, these all belong to the old order, brought up in gilded 'yali', sheltered from the wind of egalitarianism that blows, stinking of coal smoke, from the West. Among the groups here I never see anything but pleasant faces full of passion and youth.

These are men who in the daytime go about in European attire, yet, in the evening in their inviolable home setting, they will go back to wearing the silk tunic and the long cashmere fur-lined caftan. The grey jacket-coat was only a graceless temporary disguise that, to their Asiatic way of thinking, was not at all becoming.

...The fragrant smoke traces intricate and ever-changing curves in the warm air; voices are kept low, and the talk is often of the war, of Ignatieff and the worrisome Russians, the 'Moscovs', and of the disastrous fates that Allah is preparing for the Kkalif and for Islam.

The little cups of Arabian coffee have been filled and emptied several times; the wives from the harem, yearning to be noticed, undertake personally the transfer of silver trays back and forth at the half-open door. A glimpse of their fingertips, or maybe an eye or an arm, furtively withdrawn; nothing more, and at the fifth hour (ten o'clock) the door from the 'haremlike' is shut and the ladies don't appear again.

The white wine of Ismidt, which the Koran does not forbid, is poured into a single glass and, following custom, this is passed round for everyone to drink from.

The sips taken wouldn't even satisfy a girl, and the wine itself has no bearing whatsoever on what invariably follows.

Gradually heads begin to nod and the vaguest thoughts going round in them merge into some wavering dream. Against the luxurious surroundings and the gleaming mother-of-pearl, the rising smoke becomes blurred, eyes close. Gently the senses fade and oblivion, craved by every human creature, now reigns.

The servants bring in mattresses, 'yatags', on which we lie down and and go to sleep...

...Morning returns; daylight slips through the ash trellises, the painted blinds, and the silk curtains.

It's time for Izeddin-Ali's guests to freshen up, which we do in separate white marble dressing rooms that are provided with towels embroidered so richly with gold, that in England I would hardly dare use them.

We gather round the copper brazier, smoke a cigarette, and bid one another goodbye.

Rousing myself is a dreary business but I feel I've been involved in some dream from 'Thousand and one nights'. And then I find myself caught up in the morning bustle and squishing about in the mud in the streets and bazaars.

LV

All the nightly sounds I used to hear in Constantinople have lingered in my memory; they're mixed with the sound of her voice as it was whenever she imparted to me, as she often did, her weird explanations.

The most sinister sound of all was the cry of the 'beckdjis', the nightwatchmen, shouting their terrible alarm , 'yangun vâr!', such a prolonged, such a mournful wail, one that was repeated, in the deep silence, throughout every quarter of Stamboul.

And then, as the next day dawned, would come the sound of the cocks with their aubade, only shortly before the prayer of the muezzins, to us a sad chant because it told us day was here and that her returning to me before the next day threw everything into question, even her life!

During one of the first nights we spent in my lonely house at Eyoub,

we could hear a noise close to us; it was right here in the staircase of the old building. It set both of us trembling. We thought we could hear a troop of jinees at our door—or was it men in turbans crawling up the worm-eaten steps with daggers and yatagans (swords) drawn? When we were together we had cause to be afraid and to tremble. The same noise came again, this time clearer and not as frightening: so much so that it could be only one thing:

'Setchan! (Mice!),' Aziyade laughingly exclaimed, totally reassured.

The fact is, the old tumbledown house was full of them and when night came they engaged in murderous pitched battles.

'Tchok setchan var senin evde, Lotim!' she would often say. (There are a great many mice in your house, Loti!)

And this is why, one fine evening, she made me a present of young

'Kédi-bey'.

Kedi-bey (Lord Cat), who was to grow later into an enormous tomcat, full of character, was scarcely a month old at the time; he was nothing but a tiny yellow ball, studded with two big green eyes; he was very fond of his food.

She'd brought him as a surprise in one of those velvet gold- embroidered satchels Turkish children use for school.

This been hers at a time in her life, when, bare-legged and veilless she used to go for very rudimentary instruction to the old turbaned schoolmaster in the village of Canlidja, on the Asiatic side of the Bosphorus. She had profited very little from this master's teaching and she wrote very badly; not that it made me any less fond of the poor faded satchel that had been with since her early childhood ...

The evening Kedi-bey was presented to me, I was further surprised to find that he'd been swaddled in a silk napkin, into which, due to his fright on the journey, had caused him to deposit all sorts of unwelcome additions.

Aziyade, who had gone to the trouble of embroidering a collar of gilded sequins for him, was distressed to see her pupil in such an embarrassing situation. The expression on the cat's face when it was unpacked was so sorrowful that Achmet and I got a fit of the giggles.

That first sight of Kedi-bey is a memory that will stay with me for the rest of my life.

LVI

'Allah illah Allah, ve Mohammed! recoul Allah' (There is no God but God, and Mahomet is his prophet!).

For centuries, every day, at the same hour, to the same chant, those identical words have rung out over my ancient house from the top of the minaret; the muezzin's shrill voice directs the psalmody to the four points of the compass with unaltering steadiness, in tones that never vary.

Those, who now are nothing more than a handful of ashes, heard it, right here, as we, who were born only yesterday, hear it now.

That sad, mournful call to leave our sleepless bed, our night of love.

Now she must leave. Hastily we say goodbye, not knowing if we'll ever see eachother again, not knowing if that day will bring some sudden exposure, or if some act of vengeance by an old man, betrayed by four wives, won't part us forever, not knowing if that day we won't witness the playing out of one of those dark dramas of the harem, against which all human justice is powerless, all practical means of rescue impossible.

Dear little Aziyade, away she'll go, got up like a woman of low class in a coarse grey woollen garment made for her here in my house; she'll walk with her trim body all bent, sometimes leaning on a stick, her face hidden behind a thick yashmak.

A caique takes her over to the crowded area of the bazaars, whence, having first, at Kadidja's house, changed back into the clothes of a judge's wife, she is able to return to her master's harem. By way of safeguarding appearances she takes back with her from her stroll some token purchases, such as flowers or ribbons.

LVII

...Achmet was looking very important and formal: the two of us were engaged on a mission that was full of mystery. He'd received his instructions from Aziyade. I, for my part, had sworn to go with him and to do whatever I was told.

At the Eyoub jetty Achmet negotiated the the price of a caique to Azar-kapou. That done he motioned me into the boat.

He said to me solemnly:

'Sit down, Loti.'

And we set off.

At Azar-kapou I had to follow Achmet down some dubious-looking alleyways that were foul, muddy, dark, sinister, where people sold tar, old pulleys and rabbit skins. We went from door to door, asking for someone known as old Dimitraki, whom we eventually found at the back of an unspeakable hovel.

The old Greek was dressed in rags, had a white beard, and looked like an old brigand.

Achmet handed him a piece of paper with Aziyade's name calligraphed upon it, and insisted on delivering, in the language of Homer, a great long speech of which I understood not a word.

The old man reached inside a grimy chest and brought out a sort of kit of small stylets, from which he appeared to select the sharpest ones--preparations that did little to reassure me.

From what I could still remember of the classics, I was able to understand his next words to Achmet:

'Show me the place.'

And Achmet opened my shirt and placed his finger on the left side of my chest, where my heart is ...

LVIII

The operation was completed without causing me much pain, and Achmet handed to the artist a ten-piastre note which had seen the inside of Aziyade's purse.

Old Dimitraki practised the unusual skill of tattooing among Greek sailors. He had a lightness of touch and a graphic precision that was very remarkable.

I came away with a little, sore, red patch on my chest, the results of

thousands of tiny scratches, which, as they scarred over, would reveal, in Turkish, in a lovely blue colour, Aziyade's name.

According to Moslem belief, this tattoo, like every other mark or blemish on my earthly body, would go with me into eternity.

LIX

LOTI to PLUMKETT.

February 1877.

'O what a beautiful night it was … Plumkett; Stamboul was a picture!

'At eight o'clock that evening I'd left the 'Deerhound'.

'When, after a long walk, I arrived in Galata, I called by 'Our Madame' to collect my friend Achmet, and the two of us made our way through the deserted Moslem quarters to Azar-kapou.

'From there we had the choice, every evening, of two routes that would take us to Eyoub.

'One was crossing the great bridge of boats to Stamboul and along footpaths through Phanar, Balat and the cemeteries; it's a direct route and an interesting one, but at night time it's also dangerous and we only choose to go that way when there are three of us—the third being our trusty Samuel.

'That evening we went by caique from the Kara-Keui bridge and had a peaceful journey home.

'There wasn't a breath of wind, not a ripple on the water, not a sound!

Stamboul was wrapped in an immense shroud of snow.

'It was an impressive sight, one associated with northern climes,   certainly not with this city of sun and blue sky

'All the hills, covered in thousands and thousands of dark houses, passed silently before our eyes that evening, merged together in an eerie and unvarying whiteness.

'Above those human anthills, buried beneath the snow, rose the imposing grey masses of mosques and the tall thin spires of the minarets.

'The moon, veiled in mists, cast its thin, blue light over all.

'Arriving at Eyoub, we saw a soft glow filtering through the window panes, the trellises, and the thick curtains; she was already there, the first one home ...

'You see, Plunket, you, with your European houses that are foolishly open to others besides yourselves, you can have no idea how good it feels to come home—that, in itself, is worth all the weariness and the dangers ...

LX

A time will come when all this dream of love will cease to be; a time will come when everything, ourselves included, will be swallowed up in deepest night; when everything around us will have vanished, even our names on our gravestones ...

There's one country I love and would like to visit: Circassia, with its sombre mountains and its great forests. This region has a hold on my imagination, put there by Aziyade: its where she first drew breath and

burst into life.

When I see Circassians going by, fierce, half-savage, wrapped in animal skins, there's something that attracts me to these strange creatures, because the blood that runs through their veins is the same as my beloved's.

She says she remembers a large lake on whose banks she thinks she was born; and a village, deep in the forest, but not its name; also a shore where she played in the open air with the other young children of the mountain folk ...

It's a natural thing, to want to roll back a loved one's past, to want to see her child's face and her face through the years; one wants to have cherished the little girl, seen her grow, held her safe in one's own arms, well away from the caresses of others, so that no one else could own her, love her, touch her, see her. One is jealous of her past, jealous of everything she gave to others; jealous of the least movements of her heart, of the slightest word others would have heard from her lips. The present hour is not enough, one has have to have all that has passed, as well as all that is to come. Here now, with hands clasping one another, breast against breast, lips pressed to lips,

would that one could feel the other entirely, exquisitely, both at the same time and merge into a single being ...

'Aziyade,' I say to her, 'tell me a few of those little tales from your childhood, and let me know more about the old schoolmaster at Canlidja.'

Aziyade smiles and searches her memory for some new story, interwoven with fresh perspectives as well as weird asides, to entertain me with. Her favourite stories are the ones in which 'hodjas' (sorcerers) usually have the leading roles, her favourite stories being the most ancient ones, that are half-faded from her mind, and serve as little more than fleeting glimpses back to her infancy.

'Your turn now, Loti,' she says when she's done, 'Go on from where you were when you were seventeen ...'

Alas! ... All that I'm telling her in Tchengiz, I've told in other languages to other young women. And everything she says to me, others before her have said the same thing! Before Aziyade came into my life, I had this deliciously nonsensical mishmash of words repeated to me by others in voices I could scarcely hear.

Under the spell of other young women, whose memory has withered in my heart, I've loved other countries, other places, and it's all gone!

When I was with an other woman she and I dreamed this dream of eternal love; we swore to worship each other on earth and to be united, while ever there was breath in our bodies; in death we would sleep in the same grave, so that the earth should receive us and our dusts be mingled eternally. And all of that is passed, erased, swept away! ... I'm still in my youth, but it's hardly a memory.

If there is an eternity, who, then am I going to live it with? Will it be with her, or you, little Aziyade?

In these unfathomable ecstasies, in these consuming raptures, who could actually disentangle that which comes from the senses from that which comes from the heart? Is it the soul's supreme striving towards heaven, or is it the blind force of nature seeking to recreate itself and live anew? It's a question that's perpetually asked, one that all mortals have asked themselves, which makes it a nonsense to go on asking.

We come near to believing in a spiritual union and eternity because we love eachother. But how many thousands have believed the same thing in the thousands of years that the earth has been peopled, how many mortals have loved eachother and, radiant with hope, have gone to their rest trusting in the deceptive mirage of death! Alas, in twenty years, even in ten, where shall we be, little Aziyade? Lying in the earth, two lots of unknown remains, our graves, doubtless hundred of leagues apart,--and who then will still remember that we loved eachother?

A time will come when nothing will remain of our dream of love. A time will come when we shall be lost in the depths of a night in which nothing of ourselves will survive, when all trace will fade, even down to our names on our gravestones.

Little Circassian girls will still be leaving their mountains for the harems of Constantinople. The mournful chant of the muezzin will still resound in the silence of winter mornings—but it will never waken us again, never.

LXI

For a long time, the possibility of my travelling to Angora, the cats' capital, had been in question.

I was granted ten days leave of absence on the understanding that, while over there, I would stay clear of anything that could necessitate the intervention of my Embassy.

The party I was in assembled at Scutari under a cloudless sky. The Dervishes, Riza-effendi and Mahmoud-effendi, and several friends of mine from Stamboul were my travelling companions, as were some Turkish ladies, some servants, and a great deal of baggage. Our picturesque caravan moved off in the sunshine down the long avenue of cypresses that traversed the great cemeteries of Scutari. It's a place of funereal majesty; from the heights one has an incomparable view of Stamboul.

LXII

The further we went into the mountains, the more the snow slowed us down. We had no chance of reaching the cats' capital under two weeks.

After three days plodding, I made the decision to take leave of my fellow travellers; I turned southwards, accompanied by Achmet, and with two chosen horses; we would visit Nicomedia and Nicea, two ancient Christian towns.

The memory I'm left with of this first part of our travels are of wild, heavily wooded country with cool springs, deep valleys, clad with green oaks, spindle trees and rhododendrons in bloom, all on a beautiful winter's day and under a light powdering of snow.

We would overnight in those nameless hovels, known as 'hane'.

Of them all the one at Mudurlu is the most memorable. It was dark by the time we arrived. We made our way up to the first floor of this smoke-blackened old 'hane' where gypsies and bear-tamers, were already asleep, all of a jumble. The ceiling of this huge, dark room was so low that we had to move around minding our heads. The fare: an enormous cooking-pot with unspeakable items swimming in a thick sauce was placed on the floor and we all sat round it. The one and only napkin, in truth several metres long, was paid out to all the diners.

Achmet declared that he preferred to die of cold outside than spend

the night in this slovenly hovel. But by the time an hour had gone by, chilled to the bone and exhausted by fatigue, we had settled down and were fast asleep.

We were up before dawn so we could go wash ourselves from head to foot in the clear water of a spring, albeit with the wind roaring around us.

LXIII

The following evening, just as night was falling, we arrived at Ismidt (Nicomedia). We were without passports, and we were arrested.

A certain pasha was obliging enough to furnish us with two passports that he devised himself, and after a good deal of parleying, we managed to avoid sleeping behind bars. Our horses, however, were seized and slept in the pound.

Ismidt is a large and tolerably civilised Turkish town, situated beside a splendid bay; its bazaars are lively and colourful. Inhabitants are forbidden to be abroad after eight o'clock in the evening even with a lantern.

I happily recall the morning we spent here. It was early spring, the sun, set against a lovely clear blue sky, was already warm. We, Achmet and I, well replenished after our peasant breakfast, and both of us in fine fettle, with our papers in order, began our ascent of Orkham-djiami. We climbed up the narrow streets that were a mass of weeds and as steep as goat tracks. Butterflies fluttered about and insects hummed, birds sang their spring songs in the warm breeze. The old oddly shaped and broken down wooden houses were painted with flowers and arabesques; storks nestled all over the roofs and were not the least bothered if their constructions prevented a number of householders from opening their windows.

From the top of Orkhan-djiami, the eye can wander across the the blue waters of the Gulf of Ismidt, the fertile plains of Asia, and the distant Olympus of Brussa bearing aloft its high snow-capped peak.

From Ismidt to Taouchandjil, from Taouchandjil to Kara-Moussar, marked the second stage of the journey, and it was where we were overtaken by rain.

LVIV

From Kara-Moussar to Nicea (Isnik), we rode through gloomy mountains as snow fell; winter had returned. This part of our journey had its moments. A certain Ismael, accompanied by three mountain brigands, 'zéibeks', armed to the teeth, had it in mind to rob us. But thanks to the unexpected appearance of some bachibozouks, it all turned out well and we arrived at Nicea, muddied but intact. I confidently presented my Ottoman passport that the Pasha in Ismidt had devised; the official, despite my halting Turkish, was taken in by my rosary and by my garb; so there I was, for all the world a genuine effendi.

At Nicea: ancient Christian sanctuaries going back to the first centuries, a church dedicated to Aya-Sophia (Sainte-Sophie), elder sister to our most ancient churches in the West. More bear-tamers to share our bedchamber.

We'd wanted to return by way of Brousse and Moudania; but, with money running short we returned to Kara-Moussar, where our last piastres went on breakfast. We took stock of our situation and the upshot was that I handed Achmet my shirt and he went off to sell it. The money we got was enough for our return, and we embarked with our hearts light, and our purse likewise.

It was a joy, seeing Stamboul reappear. In the few days that had elapsed since we left nature had put on a different look; new plants had started to grow on my roof terrace; a whole litter of puppies had recently been born on my doorstep and were beginning to yelp and wag their tails; their mother gave us a very warm welcome.

LXV

On her arrival that evening Aziyade told me how worried she'd been and how many times she'd prayed for me:

'Allah! Sélamet versen Loti!' (Allah! protect Loti!)

She'd brought me a tiny box with something weighty inside; it smelled of rose water, as did everything of hers. Her face beamed with joy as she withdrew this mysterious little object that had been very carefully hidden in her robe and placed it in my hands.

'Here you are, Loti, she said, 'bon benden sana édié.' (I'm giving you this as a present.)

It was a heavy ring made of beaten gold and engraved with her name.

For a long time she'd dreamed of giving me a ring that bore her name, one that I could take back with me to my own country. But the poor

child had no money; she lived in great comfort, in relative luxury; she'd been able to bring to my house strips of embroidered silk, cushions and other items, which she was free to do with whatever she pleased; But she was allowed money only in small amounts; all payments were at the discretion of her maid, Emineh, so it would be difficult for her purchase a ring out of the economies she could make. Then she'd thought of her own jewellery; she feared sending anything to the goldsmiths' bazaar to be sold or exchanged; she'd had to think of other ways. The items were all her own that she gave to the smith in Scutari, in confidence, to be crushed under the hammer, and it was was these items that she brought me today, transformed into an enormous, weighty, odd-shaped ring.

At her request, I swore never to take it off, and to wear it all my life.

LXVI

It was a glorious winter's morning,-- mild, as Levantine winters are.

Aziyade had left Eyoub an hour before us; she had gone down the Horn

in a grey robe and now, dressed in pink, she was coming back up the Horn to rejoin her master's harem at Mehmed-Fatih.--She was cheerful and smiling beneath her white veil, old Kadidja was sitting beside her and the two of them were sitting comfortably in the stern of their tapering caique whose bows were adorned with beads and gilding.

Achmet and I were travelling in the opposite direction. We lay on red cushions in a long caique rowed by two boatmen.

It was that time in early morning when the splendour of Constantinople was visible all around. Palaces and mosques, still rosy pink beneath the rays of the rising sun, were reflected in the tranquil depths of the Golden Horn; flocks of 'karabataks' (cormorants) displayed the oddest behaviour as they capered around the fishing boats, before disappearing head first into the cold blue water.

I don't know if it was by chance or the whim of our 'caiqdjis'

(boatmen) that our two gilded barques passed so close to eachother that our oars engaged. Our boatmen used the occasion to exchange insults:

' Dog! son of a dog! Great-grandson of a dog!' And Kadidja saw fit to send us a secret smile, displaying her long white teeth between her dark lips.

Aziyade, on the other hand, passed us without batting a eyelid.

She seemed entirely occupied watching everything the 'karabataks' were up to :

'Neh cheytan haivan!' she said to Kadidja. (What a crafty little devil!)

LXVII

*'Who knows which one of us will still be living when the season
of delight is over?*

*Be merry, be full of joy, for the season of Spring does not
linger and will not last.*

*Hear the song of the nightingale: the season of Spring is
approaching, Every copse, every grove is a cradle of joy,*

See the almond spreading its silver blossom.

Be merry, be joyful, for Spring does not linger; it will not last.'

(Extract from an old oriental poem)

... Another Spring, the almond trees are in flower and
here am I, viewing with terror each new season that drags
me on into the night; each year taking me closer to the
abyss. My God, where am I going? What lies beyond?
And who will be at my side when I drink from the dark
cup!

...

*'It's the season of joy and delight: Spring is come. Don't say
prayers for me now, O priest; time enough for that.'*

BOOK FOUR - MENE, TEKEL, UPHARSIN

I

Stamboul, 19 March 1877.

The order for me to leave came like a thunderbolt: the 'Deerhound' was recalled to Southampton. I'd moved heaven and earth in the hope of eluding this order so I could prolong my stay in Stamboul; I'd knocked on every door, even the door of Ottoman Army, which came near to letting me in.

'My dear fellow,' said the pasha in impeccable English, and with the exemplary courtesy of a high-born Turk, 'My dear fellow, do you also intend to embrace Islamism?'

'No, Excellency,' I replied, 'it would not trouble me becoming a naturalised Ottoman, or changing my name and country, but officially I shall remain Christian.'

'Good,' he said, 'I think that's better. Islamism is not obligatory and we have little liking for renegades. I believe I'm right in

saying,' the pasha continued, 'that your services will not be accepted on a temporary basis--and, in any case, your government will be opposed to that, but you may well be offered a permanent post. So be clear in your own mind whether or not you wish to remain in our country. The fact that you are very shortly to sail with your ship does present a difficulty, as we have very little time in which to deal with the matter. Your going away, of course, would allow you a longer period in which to reflect before coming to such a serious decision, and then you could return to us later. However, if you wish, I can present your request to His Majesty the Sultan this evening, and I have every reason to believe that his response will be in your favour.'

'Excellency,' I said, 'I prefer, if it's possible, that the matter be settled immediately, later and you would forget me. All I would ask of you is a grant of leave straightaway to go visit my mother.'

I begged that I might be given an hour's grace, and I went outside to go over everything in my mind.

The hour seemed short: minutes sped by like seconds and my mind was beset by a multitude of thoughts.

I walked aimlessly through the streets of the old Moslem quarter that stretches across the heights of Taxim, between Pera and Foundoucli. The

weather was dull, heavy and humid: the colours of the old wooden houses varied from deep grey and black to reddish brown; Turkish women moved along the dry pavements in their little yellow slippers, all the while keeping themselves swathed up to the eyes in scarlet or orange silks embroidered with gold. There were vantage points along the tops of the three hundred metre high slopes, offering views over the Seraglio and its gardens of dark cypresses, over Scutariet and the Bosphorus, half veiled by the blue mists.

Abandoning one's country, abandoning one's name are more serious than one thinks when they become a pressing reality and when a final decision has to be made within the hour.

Will I still love Stamboul when I have to live my whole life here?

As for England, the monotonous pace of life there, the friends who annoy me, and the ones who take everything for granted, I can leave all that behind without any regret or remorse. I'm committing myself this country at a time of supreme crisis; this coming spring it'll be the war that decides this country's fate and my own. I shall be Yuzbachi Arif; which means I'll receive as much leave as I do now in the British Navy, so I'll still be able to go over and visit those dear to me and sit under the old lime trees at Brightbury.

By heaven, yes!... why not a 'yuzbachi' (army captain) and be Turkish for good, and remain with her ...

And at the magic moment, I could see myself: returning to Eyoub on a beautiful day, in my 'yuzbachi' uniform, and telling her I would never leave her again.

When the hour was up, my decision had been made, and it was irrevocable: it would break my heart to go away and leave her. I was granted a further interview with the pasha. I could give him my solemn 'yes' and that would bind me to Turkey forever; I would beg him to have my request put before the Sultan that very evening.

II

Once in front of the pasha, I could feel myself trembling, and a cloud passed before my eyes:

'I thank you, Excellency,' I said, 'I cannot accept. Only, please do not forget me; when I get back to England, perhaps I'll write to you ...'

III

All things considered, it was time to think of departure.

That same evening I hurried, not wanting to be detained, all over Pera,

delivering P.P.C. cards from one door to another.

Achmet, in ceremonial attire, and carrying my overcoat, followed three steps behind:

'Ah!' he said, 'ah! Loti, I see now: you're leaving us and you're paying your last visits--I've worked it out. Look, if its us you really like, and those others bore you, if you don't feel cut out to live as they do, then leave them be. Get rid of those black clothes of yours, which are ugly, and that hat, which is comical, and and come with us, now, this minute, and let all the others go hang.'

As a result of Achmet's little speech, a number of deliveries failed to materialise.

IV

Stamboul, 20 March 1877.

One last drive with Samuel. All our minutes are counted. Inexorable Time sweeps these last hours away, then we part forever!--last hours with Samuel meant wintry hours, cold and grey together with sudden March squalls.

Our agreement was that he should embark for his own country before I sailed for England. He asked, as a last favour, if he and I might drive around in an open carriage until we heard the warning whistle from the packetboat.

Samuel was greatly upset at the thought that this Achmet had taken his,

Samuel's, place, and would in due time follow me to England. The poor lad couldn't understand that there was a colossal difference between his own agonised affection and the light open, fraternal affection of Mihran-Achmet; and that Samuel was a hothouse plant that could not possibly adapt to life over there under my placid roof.

The 'arabahdji' livened things up by putting his horses into a fast trot. Samuel was wrapped like a pasha in the fur coat I was leaving him; his handsome face was pale and sad; uttering not a sound he watched as the different quarters of Stamboul went past: huge, empty squares where moss and grass were beginning to show; gigantic minarets; old, neglected mosques, white against the grey sky; venerable monuments characterised by their antiquity and decay, falling to pieces like Islam itself.

In the last of the winter winds Stamboul looked desolate and forsaken; muezzins chanted the three o'clock prayers; it was time to go.

I was very fond of my sad little friend, Samuel. I assured him, the way one assures a child, that for his sake too I would return, and that I would visit him in Salonika; but, I could tell, he knew he would never see me again, and his tears touched my heart.

V

21 March.

Poor darling little Aziyade! I hadn't the courage to tell her directly: 'I'm going away the day after tomorrow.'

It was evening by the time I got back to the house. My room was all lit up with the beautiful red rays of the setting sun; spring was in the air. The 'cafedjis' had set tables outside as they do on a summer's day; all the men of the neighbourhood were sitting in the street, smoking their 'narguilhes' beneath almond trees that were white over with blossom.

Achmet was privy to my departure. He and I in turn tried desparately to start a conversation, but Aziyade had half guessed the situation and had turned her big questioning eyes from one to the other; night began to fall, and still all three of us remained as silent as the dead.

At one o'clock Turkish time (seven o'clock) Achmet brought in an old packing case which, turned upside down, served as a table for our meagre supper. (After settling with Isaac the Jew, our cupboard was bare and we were penniless.)

Normally, having a meal together, just the two of us, was a cheerful occasion, and we could always see the funny side of our penury: there we were, two people who often dressed in silk and gold and sat on

Turkish rugs, now eating dry bread from the bottom of an old packing case.

Aziyade had sat down with me but her portion lay untouched in front of her; her eyes never left me, nor did the strange fixity of her gaze, and each of us feared breaking the silence.

'Come now, Loti, I understand,' she said. 'This is the last time, is it not so?'

And her quickened tears fell on her dry crusts of bread.

'No, Aziyade, no, my darling! There's still tomorrow. That I can swear to. After that, I can't say...'

Achmet could see that the supper had served no purpose. Without saying a word he took away the old packing case and then withdrew, leaving us in the growing dark...

VI

The next day was the day when everything was to be dismantled and taken away from this dear little house that had been furnished lovingly, bit by bit, and where every item brought back a memory to us.

I had hired a couple of labourers to see to the work. They stood there awaiting my orders. I took it into my head to send them early for their

morning meal; it gave me more time and delayed the actual destruction.

'Loti,' said Achmet, 'Why don't you make a drawing of your room? Years from now, when old age is upon you, you'll look at it and you'll remember us.'

So I used this last hour making a drawing of my Turkish room. The years will find it hard trying to take away the spell of the memories.

When Aziyade came, she found the walls bare and everything in confusion; it was the beginning of the end. Nothing but packing cases, bundles and mess; features she had been fond of had been destroyed forever. The white mats that had covered the floorboards, the rugs where only bare feet had trod, had been carted off by the Jews; everything had reverted to a dreary, wretched state.

Aziyade stepped inside almost cheerfully, somehow managing to keep up a front; but the sight of that room now it was stripped of everything was more than she could bear and she burst into tears.

VII

Like a prisoner condemned to die she asked me to grant her the favour of complying with her wishes on this, the last day.

'Today, Loti, do all I ask. Don't deny me anything. There are several things I have in mind. You're to say nothing and to approve of everything.'

Returning by caique from Galata at nine o'clock that evening, I heard an unholy din coming from my house: songs and music, the like of which I'd never heard.

In the part of the house that had recently been burned, amid swirling dust, a line of frenzied dancers were performing one of those dances

which finish only when everyone is completely exhausted; a mixed bunch of Greek and Moslem sailors, gathered from the waterfront of the Golden Horn, were dancing furiously, all the while refreshing themselves with raki, mastic, and coffee.

Suleiman, old Riza, the Dervishes Hassan and Mahmoud, who were regular visitors to our home, looked on in utter amazement.

The music was coming from my room and I found Aziyade there, engaged in turning the crank handle of one of those big, deafening, Levantine barrel organs which play Turkish dance music in strident fashion to the tinkling accompaniment of little Chinese bells.

Aziyade had removed her veil, which meant that through the half-open door the dancers would be able to catch sight of her face. It was contrary to all custom, and demonstrated a lack of even the most elementary discretion. Never before in this holy quarter of Eyoub had there been such a scandalous sight; and if Achmet hadn't declared to everyone there that she was Armenian, she would have been lost.

Achmet, was seated in a corner, withdrawn and philosophical; the whole affair was as comical as it was appalling. For a moment I felt like laughing, but when I caught the expression on Aziyade's face it wrung my heart. These poor little girls, who grow up motherless and fatherless in the shadow of the harems, can be forgiven all the preposterous notions they get in their heads, and their actions cannot be judged by the laws that govern Christian women.

She worked the barrel-organ like a woman possessed and got from its great bulk a welter of outlandish sounds.

Turkish music has been described as 'fits of ear-splitting gaiety'; that evening I understood only too well the paradox in those words.

Soon, alarmed at what she was doing and the din she was making, and deeply ashamed at finding herself unveiled in front of all these men, she went across to to a large divan, the only piece of furniture left in the house, and sat herself down; then, after ordering the organ-grinder to resume his task,she did what others were doing and asked for a cigarette and a cup of coffee.

In time-honoured fashion Aziyade's coffee was brought to her in a blue cup, about half the size of an egg and standing on a copper base.

She seemed to have calmed down; she looked at me with a smile; her sad, limpid eyes asked me to excuse the mob of people there and the uproar. Like a child, aware that it has done some foolish things, yet knowing itself to be cherished, she asked forgiveness with her eyes, which possessed more charm and persuasion than any human words could convey.

She was wearing a dress that evening which brought out a strange beauty in her. The oriental richness of her costume contrasted with the present appearance of our dwelling, now that it had reverted to a dark , squalid state. She wore one of those long-tailed basques

whose original style isn't always adhered to by Turkish women today.

The jacket was of violet silk embroidered with gold roses. Her yellow silk trousers reached down almost to her ankles and to her little feet in their gilded slippers. Her chemise of Brussels muslin, lamed with silver threads, revealed the dark-amber colour of her arms that were scented with attar of roses. Her brown hair was divided into eight plaits, each so thick that a pair of them would have gladdened the heart of any fine Parisian lady; these were spread out beside her on the divan, knotted at the end with yellow ribbons and interwoven with threads of gold, the way Armenian women have them. A mass of shorter and more rebellious hair formed a glowing halo of pale gold round the curves of her cheeks. Shades of amber deepened about her eyelids, and her eyebrows, normally close, were, this evening, joined together, conveying an expression of profound sorrow.

Her eyes were now downcast, and I could easily imagine, beneath her eyelashes, her pupils, now large and dull, angled towards the floor; her teeth were clenched and one of her red lips twitched nervously, as often happened. This contraction, which might have been regrettable in another woman, only rendered her still more engaging; in her case it indicated sorrow, or that her thoughts were elsewhere; it also revealed two even rows of tiny white pearls--that a man would have sold his soul to kiss--as well as the flesh exposed by the shortened lip that could have been the pulp of a ripe cherry.

I looked at my mistress in admiration; in this last hour I was determined to allow her beloved features to imprint themselves on my memory. The piercing music, the aromas coming from the 'narguilhes' gradually led me into a state of intoxication, that mild oriental intoxication that annihilates what has gone before, allowing one to forget the dark hours in one's life.

And such was the crazy dream that filled my mind: forget everything else and stay with her until you feel the cold touch of disenchantment or death ...

IX

Amid all the din I heard the light cracking sound of breaking china:

Aziyade hadn't moved, it appeared that in clenching her hand she'd broken her cup and the pieces had dropped onto the floor.

No great harm had been done; the thick coffee had left a mess on her fingers before spilling over onto the floor, but as far as one could tell the incident seemed to have gone unnoticed.

However, the patch on the floor was obviously spreading and a dark liquid was still trickling from her clenched hand, drop by drop to begin with then a thin black ribbon. There being only one lantern in the room it was difficult to see. I went to take a closer look: there was a pool of blood beside her. The broken porcelain had left a frightful gash in her hand, right down to the bone.

My darling's blood went on flowing for half an hour before any means of stopping it was found.

Then bowls of water, red water, started to be carried away; her hand was being held in cold water with the edges of the wound pressed tightly together: nothing could stop the flow, and Aziyade, pale as a young girl's corpse, had sunk back and closed her eyes.

Achmet had raced off to waken an old woman with the face of a witch, and it was she who finally staunched the wound, using herbs and ashes.

Having advised Aziyade to to keep her arm upright all night, the old woman demanded a fee of thirty piastres, made a few signs over the wound, and disappeared.

There was no option but to dismiss everyone and get the injured girl into a bed. For a brief while she was as cold as a marble statue and totally unconscious.

That night there was no sleep for either of us.

I could feel her suffering; her whole body was tense with pain. The injured arm needed to be kept vertical, as the old hag had recommended, and that way the pain did lessen. I myself supported her bare arm and I could feel the heat of the fever. All the fibres trembled and quivered; and I could feel them stop abruptly at the deep, gaping cut; it seemed as if my own flesh had been cut to the bone and not hers.

The moon shed its light on bare walls, bare floor, an empty room. With the furniture gone and the rough plank tables stripped of their silk cloths, the picture was one of wretchedness, cold, loneliness; outside the dogs were howling in that mournful manner, which in Turkey, as in France, is said to be a forewarning of death; the wind whistled at our door, or heaved softly like an old man in his last hour.

It hurt me to see her so resigned and in such deep despair; her dejection could have melted hearts of stone. I was everything to her, the only man she had loved and the only man who had loved her, and I was about to leave her, never to return.

'Forgive me, Loti,' she said, 'for cutting my hand and giving you all this trouble. I'm keeping you from your rest. Do go to sleep, Loti; the pain is nothing now that I know I've reached the end.'

'Listen,' I said to her, ' dearest Aziyade, do you wish me to return?...'

X

A moment later we were sitting on the edge of the bed; I was still supporting her injured arm as well as her weary head. Using the accepted phrasing of Moslem oaths, I swore to return.

'If you've married, Loti,' she said, 'it won't matter. I'll

no longer be your mistress, I'll be your sister. So marry, Loti, that's not the

main thing. I love your soul more. Just to see you again is all I ask of Allah. Then I'll be almost happy again. I'll live in that hope of seeing you; all is not ended for Aziyade.'

Shortly afterwards she was able to drift off to sleep; day was just breaking and, as I always did before the sun came up, I slipped from her side, leaving her in a deep, peaceful slumber.

XI

23 March.

I went on board and then returned in all haste. Time taken: three hours. I announced to Aziyade that departure had been put off for two days.

Two days: not much when they're the last two days life can offer you, so you must needs hurry and enjoy eachother as if you were going to die.

The news of my departure had already got around and I received several farewell visits from my Stamboul neighbours. Aziyade had shut herself away in Samuel's room and I could hear her weeping. The visitors too caught the sounds but Aziyade's frequent presence in my house had leaked out into the neighbourhood and tacitly accepted. Anyway, Achmet had announced to people the night before that she was Armenian; this assurance, coming from a Moslem, protected her.

'We always expected to see you to disappear like this,' said the Dervish, Hassan-effendi,'through a trapdoor or at the waive of a wand. Before you go, you will tell us, won't you, who you are and what brought you to live among us?'

Hassan-effendi was a man of integrity; although he and his friends would have loved to know who I was, they still hadn't the least notion because they never spied on me. Your French commissar of police would know all about you within three hours, but in Turkey you're free to get on with your own life in peace.

I let Hassan-effendi know my other names and my profession, and we promised to write to eachother.

Aziyade had been crying for hours, but her tears were less bitter. The prospect of seeing me again was beginning to take hold in her mind, and this had calmed her. She was beginning to use the phrase: 'When you come back ...'

'I don't know if you will come back, Loti,' she would say, 'only Allah knows that! Every day I will repeat: 'Allah! sélamet versen Loti!' (Allah! protect Loti!) and Allah will then do according to his will. 'But, Loti,' she continued earnestly, 'how could I wait a whole year for you? How would it be possible when I don't even know that I could wait one day, not even one hour, without seeing you? You never knew that when you were on duty I used to walk about on the heights of Taxim and that I would stay as long as I could at my mother Behidje's because from her house I could see the 'Deerhound' in the distance. Loti, you can see how impossible it is, and that if you do come back Aziyade will be dead ...'

XII

It will be Achmet's task to forward Aziyade's letters to me and to see that my letters reach her through Kadidja; what I need is a stock of envelopes addressed to him.

Now, Achmet can't write, nor can anyone else in his family; Aziyade's writing is too bad for the post. So here we are, the three of us, sitting in the tent of the public scribe, forming a typically oriental vignette.

Achmet's address is very complicated and runs into eight lines:

'To Achmet, son of Ibrahim,

who lives at Yedi-Koule,

in a cross-street that leads to

Arabahdjilar-Malessi,beside the mosque.

It is the third house after a 'tutundji'

And next door to an old Armenian woman

who sells remedies;

and a Dervish lives opposite.'

Aziyade asked for eight similar envelopes to be made, for which she paid eight silver piastres of her own money; that done, I, for my part, had to swear that I would would use them.

Tears, which she tried to hide beneath her yashmak, welled up in her eyes; my oath did not reassure her. For a start, who could believe that a piece of paper, sent all on its own from a country far away, could ever reach her? And besides, she knew very well that before long, 'Aziyade will be forgotten forever'.

XIII

That evening we went by caique back up the Golden Horn; never before had we roamed about together so freely in boad daylight. She seemed to have thrown caution to the wind, as if, for her, everything had come to an end and the world meant nothing to her.

At the Oun-Capan jetty we got into a caique. Daylight was fading; the sun was sinking behind a thundery sky.

Seldom in Europe does one see skies in such dark commotion; to the north was one of those terrible arched cloud structures that look so doom-laden, and which, in Africa, give warning of mighty storms.

'Look,' I said to Aziyade, 'that's the very sort of sky I would see every evening in the black man's country where I spent a year with the brother I lost.'

In the opposite direction, Stamboul, with its needle-like towers was fringed by a huge gash bursting with deep shades of yellow which gave out a fantastical light that verged on the ominous.

Suddenly an awful wind got up on the Golden Horn; night was closing in and we were chilled to the bone.

Aziyade's large eyes were fixed on mine in a strangely penetrating gaze; her pupils seemed to be dilating in the failing light; they seemed to read me down to the bottom of my soul. I'd never before encountered such a look from her and I was moved in a way that was unfamiliar to me; it was as if she had entered the innermost recesses of my being, and had examined them with a scalpel. The way she looked at me in the last hours we had together seemed to be asking these supreme questions:

'Who are you, you whom I've loved so much? Am I to be forgotten like a chance mistress? Or do you really love me? Is what you say the truth, and you'll come back to me?'

I close my eyes and I'm still able to recapture that expression on her face; I see her white veiled head, hardly discernible beneath the muslin folds of the yashmack: and behind her, silhouetted against the stormy sky . . .

XIV

We landed one more time down by the little square in Eyoub, which, after this evening, I would never see again.

It had been our wish to take one last look at the house where we'd lived.

The doorway was jammed with packing cases and bundles, and it was already getting dark. In one corner Achmet discovered an old lantern which he shone round our empty room. I was eager to be away; I caught hold of Aziyade's hand and led her outside.

The sky was still unusually dark; there was threat of a downpour; the houses and streets, though dark themselves, stood out clearly against the sky. The deserted street was swept by gusts of wind that made everything shake and rattle. Two Turkish women, huddled together in a doorway, looked inquisitively at us. I turned my head for one more look at the house I would never re-enter; I had a last glimpse of that little corner of the earth where I had found a little happiness ...

XV

We crossed the little square by the mosque and embarked once more. A caique took us to Azar-kapou, and from there we would make our way to

Galata, and then Top-hane, Foundoucli, and the 'Deerhound'.

It had been Aziyade's wish to accompany me and she had made a solemn promise to be brave; at that late stage she was unexpectedly calm.

We passed through Galata's noisy confusion. We had never before been seen together in these European quarters. 'our Madame' was at her door, watching for us. Seeing this veiled young woman by my side was all she needed to solve the mystery that had so long puzzled her.

We went through Top-hané and dropped down into the deserted quarters of Sali-Bazar, and then along the broad avenues that run alongside the great harems.

Finally we arrived in Foundoucli, where we were to say our farewells.

A carriage, ordered by Achmet was already there waiting to take Aziyade back to her home.

Foundoucli remains a corner of old Turkey that looks as if its been plucked from deepest Stamboul: it has its own little paved square; its by the sea; it has its own ancient mosque and golden crescent, and is surrounded by tombs of Dervishes and by sombre retreats of Ulemas.

The storm passed on and everywhere looked splendid; nothing could be heard except the far-off yapping of dogs roaming about in the silence of the evening.

The eight o'clock bells sounded on the 'Deerhound'. It was time for me to go. One short whistle told me that one of the shuttle boats was on its way to fetch me. Yes, there it was, drawing away from the black hulk of the ship and coming slowly towards us. A sad time--a time when farewells cannot falter.

I kissed her lips and her hands. Her hands trembled slightly, but beyond that she was as calm as I was; her skin felt as cold as ice.

The boat arrived. She and Achmet withdrew to a dark corner of the mosque; we pushed off and I lost sight of them.

A moment later I heard the brisk grind of wheels as the carriage bore my beloved away forever!... a grim sound like that of earth tumbling into a loved one's grave.

Truly, an end; there's no turning back! If I ever return, as I have sworn to do, the years will have scattered their own dust on the past, or I will have dug a gulf between the two of us by marrying someone else, and she will never more be mine.

I was seized by a mad urge to run after the carriage

and lock her in my arms, so we could go on loving each other with all the strength our souls could provide, till death made us part.

XVI

24 March.

A rainy March morning. An old Jew is taking away the furniture from

Arif's house. A glum-faced Achmet watches on.

'Achmet, where is your master going?' ask the early-bird neighbours from their doorways.

'I've no idea,' replies Achmet.

Damp packing cases and bundles soaked through by the rain move off down the Golden Horn in a caique *du côté de lamer.*

And that's the last of Arif, the role as ceased to exist.

The whole oriental dream is ended; this particular stage of my life--the last, no doubt, to enchant me--is gone for good. And time may, perhaps sweep away even the memory of it.

XVII

When Achmet came on board with the items of luggage he was supervising,

I was able to let him know that we'd been granted a

further respite of at least twenty-four hours. A storm was coming in off the Marmara.

'So let's have another look round Stamboul,' I said; 'Call it a posthumous stroll; it won't be without a certain melancholy charm. But not with her; I shan't ever see her again.'

I went to 'our Madame' to change out of my European clothes; and once more Arif-effendi in person stepped out of the tavern and crossed over the bridges, rosary in hand, displaying the solemn air and proper bearing, assumed by all good Moslems who take themselves seriously, as they proceed piously to their prayers. Achmet, who walked beside Arif-effendi, had put on his finest raiment. He had asked me to leave arrangements for this last day to him; then, for a short while, he lapsed into a silent grieving.

XVIII

When we'd done the rounds of all our familiar haunts, smoked a goodly number of hookahs and paused by every mosque , we found ouselves in the evening back in Eyoub, drawn yet one more time to the place where I was now no more than a shelterless stranger, soon to be entirely forgotten.

My entrance into Suleiman's cafe caused a sensation: they'd all thought I'd disappeared like a character in a play, gone for good and all.

There was a large motley crowd there that evening;, many faces were entirely new to me--goodness knows where they were from; they were the sort of people you'd expect in a beggars' den, or the nearest thing to one.

Achmet, however, set about organising a farewell celebration for me and sent for an orchestra: two oboes, that were as shrill as bagpipes, a barrel organ, and a big drum.

I agreed to all of this on the solemn promise that there would be no breakages and no blood.

We were going to be senseless before the evening was out; for my part, I asked for nothing better.

They brought me my hookah and my cup of Turkish coffee that a child would replenish every quarter of an hour. Achmet got everyone to make a circle and hold hands, and then he invited them to dance. A long chain of odd-looking figures, caught by the unsteady light from the lanterns, started to sway about in front of my eyes. The deafening sound of the music made the roof beams of this hell-hole tremble; copper utensils, hanging from the black walls emitted metallic vibrations; the oboes strained to produce their shrill notes and the ear-piercing merriment burst into a frenzy.

After an hour of this all those in the room were intoxicated with the dancing and the din; it was a perfect way to celebrate.

I could see only what my cloudy vision allowed me to see; my head was

becoming a mass of strange and incoherent thoughts. Time after time , groups of dancers, exhausted and gasping came and went. The whirling didn't stop and after every full turn Achmet would break a pane of glass with the back of his hand.

One by one the cafe windows clattered onto the floor where the pieces were crushed beneath the dancers' feet; Achmet's hands were a mass of deep cuts which bled onto the floor.

It would seem that noise and blood were requisite accompaniments to

Turkish grieving.

I was disgusted by the whole celebration and I was worried too for the future now that I'd seen Achmet behave so stupidly and be so unconcerned by the promises he'd made.

I got up to leave; Achmet understood and followed me in silence. The cold outside restored our calm and composure.

'Loti,' said Achmet,'where are you going?'

'Back on board,' I replied, 'I want no more to do with you. I'll keep my promises in the same way you kept yours this evening; you won't be seeing me ever again.'

I walked further along and had a word with a boatman out late concerning the fare to Galata.

'Loti,' Achmet called, ' forgive me; you can't part from your brother like this.'

And then came tears and entreaties.

Not that I wanted to part from him in that way, but I'd already decided that what was needed was penitence on his part and a warning from me. I wasn't going to relent.

He tried to hold me back with his bloodstained hands, then he clung to me in despair. I pushed him violently away, causing him to fall against a pile of wood that collapsed with a crash. Some bachibozouks de

patrouille , thinking we must be criminals came towards us, shining a lantern.

We were at the water's edge in a deserted part of the suburb, far from Stamboul, and Achmet's bloodied hands would appear to confirm that all was not well.

'It's nothing,' I said to them,' The young fellow's has been drinking and I'm taking him home.'

With that I took hold of Achmet's hand and went along with him to where his sister, Eriknaz, lived. When she'd finished bandaging his fingers, she gave him a lecture and packed him off to bed.

XIX

26 March.

The last postponement of my departure. One more day.

One more day, one more changing clothes at Bizum Madame's, and being back in Stamboul.

The weather was dull and thundery, the breeze warm and mild. We spent two hours smoking our ng's beneath the Moorish arcades in Sultan-Selim Street, where white colonnades, misshapen by the years, alternated with funerary pavilions and lines of tombs. Branches of trees, all pink with blossom, rose above the grey city walls; new growth was showing everywhere and spreading merrily across the sacred marble slabs.

I adore this country and all details like these delight my fancy; I love it because its hers and she brought it to life with her presence--she, who is still here, close to me, and whom I shall never see again.

The setting sun finds Achmet and me sitting in front

of the Mehmed-Fatih mosque on a favourite seat where we've whiled away many an hour.

Here and there across the immense square groups of Muslims smoke and talk, enjoying, as a matter of course, the enchantment of the spring evening.

The sky has resumed its calm and there's not a cloud to be seen. I love this place. I love the oriental life. I can scarce believe it has come to an end and that I'm about to leave.

I'm looking at that black portico down below, and at that empty street which loses itself in a gloomy and unattractive part of the city. That is where she lives; were I to take a few steps forward I would see her house again.

Achmet has been following my gaze. He gives me a very worried look. He's guessed what I'm thinking and knows what I have in my mind.

'Ah!' he says, 'Loti, if you love her, have pity on her. You've said your farewell; time now to leave her in peace.'

But I'm determined on seeing her again; I can't help myself.

Achmet implores me with tears in his eyes to see reason, or just plain common sense. Abeddin is there, old Abeddin, her master, and it's madness to make any attempt to see her.

'Besides,' he says, 'even if she got out of the house, there's nowhere you can take her. Loti, in the whole of Stamboul, where could you find lodging for you and another man's wife? If she sees you, or if the other wives tell her you're outside, she'll be like someone gone crazy and she'll be ruined, and then tomorrow you'll abandon her in the street. What is it to you? You're going away.

But if you do this thing, then I'll hate you because it means you're heartless.'

Achmet hangs his head and begins tapping the ground with his foot, which he's in the habit of doing whenever my will proves stronger than his.

I let him be, and make for the portico.

I lean back against one of the pillars and look down onto the dark, deserted street, which might have been in a city long abandoned.

Not a window open, not a soul about, not a sound to be heard; just the sight of new grass coming up between the cobbles; and the dried up carcasses of a couple of dogs lying on the paving.

It used to be an aristocratic neighbourhood. The old, dark-timbered mansions point to a life of quiet opulence; balconies are shuttered and ‚shaks' , like great prows, jut far out over the street; behind the shrewdly positioned iron grills can be seen ash trellises, which in former days have been painted with trees and birds. All windows in Stamboul have been similarly painted and firmly shut.

In western European towns and cities, life inside a home can easily be guessed at from the outside; when the curtains are drawn back passers-by will be able to catch a glimpse of human faces, young or old, ordinary or refined.

But there's no peering into a Turkish home. Should the door be opened to admit a visitor, it will only be half-opened: there'll be someone behind the door who will close it as soon as possible. One never has any inkling of what's inside.

That large house over there, painted a dark red, that's Aziyade's.   Above the door is a sun, a star, and a crescent; every plank is wormeaten. The trellis-work on the 'shaks' have an assortment of blue tulips and yellow butterflies painted on them. Nothing stirs, nothing to indicate that there's anyone living there. One never can tell with Turkish windows whether or not there's someone behind watching you.

The square on the higher ground behind me is now golden from the rays of the setting sun; here in the street everything is already in shadow.

Half-hidden behind a section of the city wall I gaze down at the house and my heart beats wildly.

I think back to the day I saw her, saw her for the first time in my life, through the grill of the house in Salonika. I no longer know what I want, nor what I've come looking for; I'm afraid the other wives may

laugh at me, afraid of looking ridiculous; more than anything I'm afraid of being the ruin of her ...

XX

When I got back to Mehmed-Fatih square, the whole of the enormous mosque, the Arab porticoes, and the towering minarets were bathed in the golden splendour of sunset. The Ulemas, emerging from evening prayer, had all paused around the entrance and down the great flight of steps; they were looking into the sun. In the middle of the throng a young man pointed to the sky, a young man with a wonderfully mystical face. The white turban of the Ulema was wound round his fine broad forehead; his face was pale, his beard and his large eyes were as black as ebony.

He was pointing upwards towards some invisible thing and gazing ecstatically into the depths of the blue sky; then he called:

'See, there is God. Look, everyone. Behold Allah! Behold the Eternal !

And, like the rest, Achmet and I rushed to get close to the Uleman who beheld Allah.

XXI

Alas, we didn't behold a thing. It would certainly have been a blessing if we had. Then, or at any time, I would have given my life for that divine vision; yes, my life for just one sign in the sky, one manifestation of the supernatural.

'He's lying,' said Achmet. 'What man has ever seen Allah?'

'Ah! It's you, Loti,' exclaimed Izzet, one of the Ulema. 'You as well, do you also want to behold Allah?' He added with a smile: 'Allah doesn't reveal himself to infidels.'

'He's mad,' said the Dervishes.

We saw the visionary being led back to his cell.

Achmet turned this diversion to good account by taking me up the hill overlooking Marmara; it was the furthest I could be from Aziyade. When night fell we had difficulty finding our way.

XXII

We dined beneath the arcades in Sultan-Sélim Street. It was already late for Stamboul; Turks go to bed with the sun.

One by one the stars appeared in the clear sky; the moon shed its light

on the wide, empty street, on the Arab colonnades and on the ancient tombs. Here and there the Turkish cafes that were still open cast a red glow onto the grey pavements; what few passers-by there were carried lanterns; whichever way one looked dismal little lights marked the location of a funerary pavilion. I took a last look at these familiar scenes; this time tomorrow I would be far from this land.

'We'll go down as far as Oun-Capan,' said Achmet, who again had my full permission that evening to plan what we were to do next. 'We'll hire horses to take us to Balate, then a caique to Pri-pacha, and then we'll go stay the night at my sister's, Eriknaz; she's expecting us.'

We got lost on our way to Oun-Capan, and our lanterns set all the dogs barking; we knew our Stamboul well, but even old Turks find themselves

lost at night in that maze of streets. There was no one about to give us directions; always the same little streets, uphill, downhill, needlessly twisting and turning like paths in a labyrinth.

At Oun-Capan, just outside Phanar, two horses were ready, waiting.

A runner went ahead of us carrying a lantern at the

end of a two metre pole; we took off like the wind.

Phanar, gloomy and stretching forever, was asleep; silence reigned. Even at high noon the sun was doubtless loath to illuminate the dark, narrow streets we rode down. There was hardly enough room for us to ride abreast. On one side was the high city wall, on the other side

tall houses, older than Islam and half-covered in ironwork, their upper storeys extending in an arch over the humid alleyway. A rider needs to duck to ride safely beneath the balconies of these Byzantine houses on account of the massive stone supports hidden in the impenetrable darkness.

It's the way we used to come every evening when we were returning to our home in Eyoub; at Balate we knew we were close to home; but now that same home has ceased to exist ...

We roused a boatman to ferry us across to the opposite bank in his caique ...

It's open country over there. Great black cypresses tower above the plane trees.

Lanterns in hand, we began to climb the paths leading to Eriknaz's house.

XIII

Eriknaz-hanum's pleasant, ordinary face was nevertheless distinctive; her skin was pale and waxen, her eyes and her eyebrows were as black as a raven's wing. She received us unveiled, like a Frankish woman.

The whole interior of her house spoke of orderliness, comfort and total cleanliness. When we arrived two of her friends, Murrah and Fenzile, who had been keeping her company, hid their faces and took flight. They had been sewing gold sequins onto little red slippers that turned up at the toes like horns.

As usual my little friend, Alemshah, Eriknaz's daughter and Achmet's niece, made herself comfortable on my knees and dropped off to sleep; she's a pretty little creature, three years old with great jet-black eyes, and is sweet and dainty as a doll.

Following coffee and a cigarette, we were brought two mattresses, two 'yatags', and two coverlets, all as white as snow. Eriknaz and Alemshah

bade us goodnight and retired. As for us, we both fell into a deep, deep sleep.

Next morning we were woken by brilliant sunshine. Off we went down the paths that lead to the Golden Horn. An early morning caique was there waiting for us.

The innumerable black houses of Pri-pacha, that rose together pyramid-like in tiers were basking in an orange light that sparkled at every window. From the distance Eriknaz and Alemshah, perched on their rooftop, their robes red in the ascending sun, watched us depart.

We went past Eyoub; there was Suleiman's cafe, the little square by the mosque, and Arif-effendi's house with the morning light full upon it. Nobody down by the water; every house still locked and everyone still asleep.

And there was my house, the house that so often looked sombre and depressing beneath the snow, buffeted by north winds. Now it was offering a final image of itself in a blaze of sunlight.

That last sunrise was one of extraordinary splendour; all along the Golden Horn from Eyoub to the Serail, domes and minarets stood out against the limpid sky in pink and iridescent tints. Gilded caiques in their hundreds were starting to travel about transporting picturesque men and veiled women.

An hour later we were on board. Everything was chaotic but this time it really was departure.

It was set for midday.

XXIV

'Come on, Loti,' said Achmet, 'let's go back to and smoke our hookahs together one last time ...'

We raced through Sali-Bazar, Tophane, Galata. We stopped at the Stamboul bridge.

A scorching sun beat down on the crowded streets. Spring had indeed come to stay, and it was arriving the very day I was leaving. The noonday brilliance streamed down on that harmonious grouping of walls, domes and minarets that crowns Stamboul; the light fell on the myriad of rainbow colours in the crowd.

Boats arrived and departed , packed with eye-catching figures; hawkers yelled at the tops of their voices as they jostled through the throngs.

We were well acquainted with every one of these boats that had taken us all over the Bosphorus; we knew every booth on Stamboul Bridge, every passer-by, even every beggar--a whole crew composed of the maimed, the blinded, the one-armed, the harelipped and the legless. Every petty crook in Turkey was out and about that day: I distributed alms to all and sundry, and reaped a litany of blessings and salaams.

We halted for a while in Stamboul in the great square of Jeni-djami, facing the mosque. For the last time in my life I was able to relish the feeling of being a Turk, and sitting beside my friend Achmet, smoking a n in this oriental setting.

Today was a real celebration of Spring: a display of costumes and colours. The whole populace was outside, sitting beneath the plane trees, around the marble fountains, in bowers of vines that were soon to be clad with tender leaves. Barbers had moved all their tackle into the street and were seeing to their customers out in the open air: pious Moslems were solemnly having their heads shaved--bar a tuft left at the top of the head for Mahomet to take hold of when he carries them off into paradise.

...Who will carry me away into whatever paradise attends?--some place that's not an extension of this old world, which I find tedious and wearisome, some place that won't keep on changing, where I shan't be separated everlastingly from what I love and what I have loved?

If someone could simply give me the Moslem faith, I would go off with tears of joy and take up the green banner of the Prophet.

(Stupid of me to mention keeping that tuft on the top the head ...)

XXV

'Loti,' said Achmet, 'tell me a little about the voyage you're going to make.'

'Achmet,' I replied, 'when I've crossed the Sea of Marmara, the Ak-Déniz (the Old Sea) as you call it, I'll cross another, much bigger, sea, to get to the country of the Greeks; and after that I'll cross a still bigger sea that will take me to the the country of the Italians, the country of 'our Madame', and then I'll go across a sea that's greater than the others, which will take me to the southern tip of Spain. If only I could stay in these marvellously deep blue waters of the Mediterranean, I wouldn't be so far from you; it would mean we would share, to some extent, the same sky; and the ships that make the regular run from the Levant would rgularly bring me news from Turkey. But finally, I'll come to an ocean that's so huge there's nothing to compare with it, and for several days I'll have to be guided by a star (the North Star) if I'm to reach my native land--a land where the weather's more often rainy than fine, where there's more cloud than there is sun.

'Over there I'll be far away from you, and this country of mine is very unlike yours; everything is paler and all the colours are duller; its like being in one of your mists, only not as transparent.

'The countryside is so flat, you've never seen the like —unless it was when you made your pilgrimage, as a good Moslem should, to Mecca, to the Tomb of the Prophet—only, instead of sand there is green grass and large fields of ploughed land. All the houses are square and similar. You look from your window and pretty well all you see is your neighbour's wall; often there are times when the flatness becomes oppressive and you's love to get higher and see farther.

'What's more, there are no steps, as there are in Turkey; and, honest truth, I once had the bright idea of stepping, for a change, onto my roof. From then on people round about thought me very odd.

'Everyone dresses the same. Grey overcoat, hat or cap; it's worse than Pera. Everything is worked out beforehand; rules are made and go by numbers; there are laws for eveything, and rules for everyone, so that the lowest wretch, or the hosier, or the barber has the same rights in everything as has an intelligent and strong-minded fellow like, for instance, you or me.

'In short, my dear Achmedim, believe it or not, if we did in England only a quarter of the things we do every day in Stamboul, we would be hauled in by the police and told to explain ourselves.'

Achmet was well able to comprehend this overall view of western civilisation; he deliberated for a moment.

'When the war's over, Loti,' he suggested, 'why not bring your family to live in Asiatic Turkey?'

'Loti,' Achmet said later, 'I want you to take this rosary, which came to me from my father, Ibrahim, and to promise me never to part with it. I know for certain that I'll never see you again. A month from now and we'll be at war. It'll mean the end of us poor Turks and the end of Stamboul; the Muscovites will destroy us all; and when you return, your friend Achmet will be dead.

'His body will lie somewhere up on the northern front. He won't even have a little grey marble stone under the cypress trees in Kassim-Pacha cemetery; Aziyade be living in the east, and you'll never find any trace of her;

there'll be no one who can talk about her with you. Loti, he said, his tears overflowing, 'Loti, stay with your brother.'

Alas, I fear the Muscovites as much as Achmet does. I shudder at the horrible thought that I could actually lose all trace of her and that I'd never in the whole wide world, find anyone who could talk with me about her!...

XXVI

Muezzins were climbing up to their minarets; the time for the midday namaze was almost here; time for me to depart.

On my way back through Galata I called on 'Bour Madame' to say farewell. I came very close to kissing the wicked old thing.

Achmet accompanied me back on board where we said our farewells amid all the hubbub of visitors and preparations for putting to sea.

We cast off, and Stamboul moved away ...

XXVII

At sea. 27 March 1877.

A pale March sun is setting over the Sea of Marmara. The air is cold and sharp. The shores, dismal and bare, are fading into the evening mist. Is this the end, God?

Will I never see her again?

Stamboul has disappeared; the tallest domes of the highest mosques are all gone from sight, wiped clean away. If I could see her just for one minute ...; I'd give my life if I could only touch her hand. I have this mad longing to have her by me.

My head is still full of all that oriental din, the crowds in Constantinople, the commotion as we got ready to sail; now I'm finding even the calmness of the sea oppressive.

If she were here I would weep—that's something I've not been able to do; I would lay my head on her knees and I'd cry like a child; she would see my tears and would know they were true. I was very calm and cool when I said goodbye to her.

And yet, I adore her. It goes beyond rapture: I love her with the tenderest and purest affection; I love her soul, I love her heart, and both of them are mine. I will still love her when youth has passed, when the delights of the senses have waned, when our unknowable future brings us old-age and death.

The stillness of the sea, the paleness of the March sky, they wring my heart. God, my suffering is real; my anguish couldn't be greater if I'd seen her die. I kiss everything I have of hers; would that I could weep, but I can't even do that.

At this hour my beloved will be in her harem, in some part of that

utterly dismal and heavily barred residence. She'll be lying there, what's left of her, uttering no words, shedding no tears, as night closes in.

Achmet had stayed, sitting on the quayside at Foundoucli, his eyes fixed on the ship. I lost sight of him and of that familiar corner of Constantinople where every evening he or Samuel would come and wait for me.

Achmet as well thinks that I shan't be returning.

My poor little friend, Achmet. I was very fond of him--I am still,--a sweet-natured, generous friend.

My thoughts are now turning away from the Orient; that dream is ended. Ahead lies my homeland. In peaceful Brightbury a glad welcome awaits

me. I too will be glad. I love my family; but what a dreary place it is.

All the same, I'll be pleased to see once more the place where I was born and spent my childhood. I can picture the old ivy-covered walls, the grey Yorkshire sky, the old roofs, the moss, and the lime trees, witnesses of yesteryear, witnesses of my earliest dreams and of the happiness that nothing in the world can ever restore to me.

I've often before returned to my family home, my heart tortured and broken. I've brought back with me any number of new passions and hopes, only to see them crushed. The house is full of painful memories; its blessed calm is no longer of benefit to me; this time I'll suffocate like a plant deprived of sun.

to LOTI, from his SISTER

Brightbury, April 1877.

'My dear brother,

'I too must write to welcome you back to your native land. Through our affection may the One in whom I trust grant a lightening of your spirits and an easing of your heartaches. So overjoyed are we at your return, I'm confident that every care possible will be taken to help you.

'I often reflect on the fact that when someone is so loved and cherished and is foremost in the thoughts and affections of so many people, he can have no cause whatsoever to think his life accursed and that he is of no account in this world. I wrote you a long letter and sent it to Constantinople, but in all likelihood you never received it. In it I let you know how much I sympathised with your difficulties, even with your sorrows.

Va, I've shed tears more than once over Aziyade's story.

'I think, dear little brother, that it's not altogether your fault if everywhere you go you leave a small part of your sad life behind. Although your life is far from being a long one it's a life that has known so much strife ... but you know it's my belief that someone will come along soon and captivate you, and that you'll find its the best thing that could have happened to you.

'The nightingale and the cuckoo, the warbler and the

swallows are already heralding your arrival; you couldn't have chosen a better time of the year. It could be we'll find a way to keep you here a while and thoroughly spoil you.

'Goodbye just for now. Kisses from us all!

XXIX

Translation of a near-unintelligible letter in Turkish, written to Achmet's dictation by a public letter-writer in Emin-Ounou at Stamboul, and addressed to Loti at Brightbury:

'ALLAH!

'My dear Loti,

'Achmet sends you many greetings.

'I forwarded your Mytilene letter to Aziyade through old Kadidja; Aziyade keeps it close to her hidden in her robe, and as yet not had it read to her because she has not been out since you left.

'We were reckless during those last days and Old Abeddin suspected and then guessed all. Kadidja said he has not taken Aziyade to task, nor got rid of her because he was very fond of her. But he never goes into her room, or speaks to her or gives her any attention. The other wives

too have deserted her, except Fenzil-hanum, who went

to consult the 'hodja' (sorceress) on her behalf.

'She has been ill ever since you left; however, the tall 'ekime' (doctor) who visited her said there was nothing wrong with her and has not been back.

'She is being looked after by that old woman who once stopped her hand from bleeding. She is her confidante and it is my belief that she was bribed to denounce her.

'Aziyade wants you to know that she has no life without you; that she cannot imagine you ever returning to Constantinople; that she cannot believe she will ever be allowed to look into your eyes again; and that for her the sun has gone.

'Loti, the words you said to me, do not forget them; the promises you made to me, never forget them! Do you really believe that I can be happy for a single moment in Constantinople without you here? I cannot, and when you went away, my heart broke with the sorrow of it.

'So far I have not been called for military service on account of my father who is very old; but I think I will be called up soon.

'I, your brother, salute you,

'Achmet.

'P.S.--Last week Fire broke out in the Phanar quarter of the city. Phanar is burnt to the ground.'

XXX

LOTI to IZEDDIN-ALI at Stamboul

Brightbury, 20 May 1877.

'My dear Izeddin-Ali,

'Here am I in my country--so different from yours!--
sitting beneath the same old lime trees that sheltered me
as a child. This little town of Brightbury that I told you
about in Stamboul is surrounded by my beloved green
oaks. It is Spring, albeit a pallid one: rain and mist--a
little bit like your winter.

'I am back to the wearing standard western dress: hat
and grey overcoat; but there are times when I feel that the
proper clothes for me are the same as yours, and that just
for a while I'm really going about in disguise.

'However, I love this little corner of my native land
and the family home that I've so often deserted. Here
people are loving towards me and I love them in return; it
was their affection that made my childhood happy and
peaceful. I love everything that surrounds me, especially
the countryside with its ancient woodland that has a
charm all of its own—a strongly pastoral charm which I
find difficult to describe--call it the charm of the past,
the very distant past, and of ancient shepherds.

'News is overtaking news, my dear effendim, news,
that is, of the war; events have gathered pace. I had hoped
that British people would take sides with Turkey, and I
feel only half-alive, being so far from Istanbul. You have
my passionate support; I love your country and I wish it

well with all my heart, and doubtless you'll soon be seeing me again.

'But then, you have already perceived my love for her whose presence you may have privately questioned but accepted nevertheless. Yours is a generous heart; you rise above all conventions and all prejudices. I can assure you, I do love her and its for her sake most of all that I will soon be returning.'

XXXI

Brightbury, May 1877.

I was sitting at Brightbury beneath the old lime trees. Above my head a blue tit was singing its complicated and very long song and putting its whole heart and soul into it; its song awoke in me a world of memories.

At first it was all confused, as if the memories were distant; then gradually the images grew clearer and more precise. Suddenly it all came back.

Yes, it was over there in Stamboul,--one of our hugely rash, indulgent days, one of our recklessly truanting days. But then Stamboul was so big and we were so little known there!... And Abeddin was far away in Andrinopolis!..

It was a lovely winter's afternoon and we were out walking together, just the two of us, happy as a children to find ourselves, as luck would have it, out in the sunshine, free to wander about the countryside.

Yet we had chosen a dreary place for our first jaunt: we kept alongside the high city wall; there couldn't be a more lonely place; not a thing seemed to have stirred since the time of the last Byzantine emperors.

The great city does all its communicating by water, and the silence surrounding its ancient walls is as complete as the one encountered when approaching a necropolis. If, here and there a gate has been built to allow passage through the thick ramparts, they are to no purpose as no one ever uses them and they may just as well be filled in. Besides, there are always those low, handy, mysterious little doors that have above them gilded inscriptions and curious adornments.

Between the inhabited section of the city and its fortifications lie vast tracts of waste ground, occupied by suspicious-looking shanties

and tumbled down remnants that date from every epoch of history.

And there's nothing beyond the walls to interrupt their own, long monotony--allowing, that is, for the white stems of minarets, that popped up at intervals; always the same battlements, the same towers, the same sombre hue, deepened over the centuries,--the same regular lines that run straight and mournful until they disappear over the furthest horizon.

We went on foot, just she and I, to the foot of the high walls. In the countryside all around were groups of gigantic cypresses, tall as cathedrals, and, in their shade, thousands of burial places crowded together. Nowhere have I seen as many cemeteries as I have in this country, nor as many tombs, nor as many dead.

'These places,' said Aziyade, 'were very dear to

Azrael's heart. It's where he would alight when night fell; he would fold his great wings and walk about like a man beneath these awesome shades.'

Everywhere was silent; the grounds were impressive and solemn.

On the other hand, both of us were cheerful, happy in our escapade, glad to be young and free, wandering, for once, wherever chance took us, pleased, like everyone else, to be out in the open and beneath a clear blue sky.

Aziyade's 'yashmak' was very thick and she'd pulled it down over her eyes to the point where all her forehead was hidden. She kept her veil tightly to her face and I could scarcely glimpse through the gap the clear, quick pupils of her eyes as they travelled about behind it. Her borrowed 'feredge' was dark in colour and severe in cut-- a style highly untypical of young and elegant women. Old Abeddin himself would not have recognised her.

We walked with swift, supple steps, brushing the modest white daisies and the short January grass, filling our lungs with the good, sharp, bracing air, typical of fine winter days.

All of a sudden, through the great silence, we heard the delightful singing of a tit, exactly like the one I heard today; birds of the same species, wherever they are in the world, repeat the same song.

Aziyade, taken aback, stood stock still; with a comical look of amazement, she pointed her henna-dyed finger directly towards the little songster, perched close to us on a cypress branch. This little bird, so small and so alone was taking such pains producing this sound; it struggled so earnestly, so joyfully that the sight of it gladdened our hearts and set us laughing.

And we remained there listening to it right up to the moment when it flew off, frightened by a line of six large camels, tied head to tail, advancing unheedingly towards it

Afterwards...afterwards we saw appear a group of women dressed in mourning and heading towards us.

They were Greek women; two of their priests, their 'popes', walked at the head of the procession; the women were carrying a little corpse that lay, its face uncovered, on a bier, as is their national custom.

'Bir guzel tchoudjouk.'('What a pretty little child!') said Aziyade,

turning serious.

So it was: a pretty little girl of four or five, a delightful wax doll that seemed to be asleep on some cushions. She was clothed in an elegant white muslin dress and wore a wreath of gilded flowers on her head.

There was a grave already dug by the side of the road. The Greeks don't care where they bury their dead; it may be along roadsides, or by a walls ...

'Let's get closer,' said Aziyade, as if she were a child again, 'they''ll give us sweets.'

Digging the grave had involved disturbing a corpse that couldn't have been in the ground all that long. The soil that had been shovelled out was full of bones and scraps of various fabrics. The most notable item was an arm, bent at a right angle, its bones still red, and held together at the elbow by something that the earth had not had time to devour.

The two 'popes' had lots of long hair, like women; they wore disturbingly grubby and faded gold vestments, and were accompanied by four odd young rascals who constituted the choir.

They muttered something over the dead child; then the mother removed the wreath of flowers and carefully tucked the little girl's blond hair inside a little night cap; such close attention to a child would have normally caused us to smile, but not there, with that mother.

When the child's body had been laid at the bottom of the grave, on the bare, damp soil, with no board or bier underneath, the same loathsome earth was shovelled over her. Everything went back into the hole onto

that pretty, waxen face, including the bits of bone and the elbow; she was buried in no time at all.

Yes, we were given sweets. I hadn't known of this Greek custom.

A girl with a large bag of sugared almonds gave a handful to each person there, including us, even though we were both thought to be Turks.

When Aziyade put her hand out to receive hers, her eyes were full of tears ...

XXXII

The little bird, beside itself with the joy of being alive, was, in point of fact, singing its heart out in the middle of a burial ground!...

BOOK FIVE - AZRAEL

I

20 May 1877.

... At last the clear sky and the blue waters of the Levant. Something was taking shape out there; the horizon was suddenly fringed with mosques and minarets;--my heart beat faster, it was Stamboul!

Setting foot ashore, I experienced a keen burst of emotion at finding myself back in this country ...

No Achmet there at his post, caracoling on his white horse at Top-Hane. Galata itself is finished; I could tell there was something terrible going on beyond the city, something like a war of extermination.

... Back in my Turkish clothes again, I made haste to Azarkapou. I climbed into the first caique that came along. The boatman recognised me.

'And what of Achmet?...,' I said.

'Gone. Gone off to war!'

I called at the home of Eriknaz, Achmet's sister.

'Yes, gone,' she said. 'He was at Batum, but since the battle we've had no news of him.'

Eriknaz's black eyebrows were drawn together by her pain. She shed bitter tears for the brother whom others had stolen from her; and little Alemshah, gazing at her mother, shed tears of her own.

I then went to Kadidja's house; but the old woman had moved and no one could tell me where to.

II

I headed for the mosque of Mehmed-Fatih, close to the mansion where Aziyade would be. I made no effort to halt whatever plan was racing through my confused mind, though I scarcely knew what I was even

thinking of doing; all I was conscious of was my need to get close to her and to see her!...

I made my way through the heaps of ruins and ashes that were once opulent Phanar but now no more than one vast scene of devastation, a long succession of gloomy streets, cluttered with black and charred debris. This was the Phanar I used to pass through so cheerfully of an evening on my way home to Eyoub where my beloved was waiting ...

There were shouts in the streets; groups of half-clothed men, levied for the war, poorly armed and half-savage, were sharpening their 'yatagans' on the stones and carrying about with them old green flags striped with white inscriptions.

I walked for a long time. I traversed the empty quarters of Eski-Stamboul.

All the time I was getting nearer. I entered the gloomy street that lead up to Mehmed-Fatih, the very street where she was living!...

In the sun, objects lying outside the house took on a somewhat sinister appearance. My heart sank. There was not a soul in the whole dreary street, only the sound of my footsteps breaking the heavy silence.

Pattering along over the paving stones and the grass, and keeping close to the wall, came the figure of an old woman. Visible beneath the folds of her robe was a pair of thin, bare, ebony coloured legs; her head drooped forward and she was talking to herself... it was Kadidja.

She recognised me. She let out an untranslatable 'Ah' in that high-pitched voice I associate with negresses or monkeys, and accompanied it with a jeering snigger.

'Aziyade?' I asked.

'Eulu! eulu!' she retorted, dwelling deliberately on the weirdly savage words that in Tartar signified death.

'Eulu! Eulmuch!' she repeated as if to someone who didn't understand.

Then, sneering with hate and satisfaction, she pursued me mercilessly with her taunt:

'Dead! Dead!... she's dead!'

Coming without any warning a word like that isn't easy to register straightaway; it takes a moment for the pain to fix its grip before it bites into the heart. I kept on walking. I was horrified at my calmness. All the while, like one of the Furies, the old woman followed me step by step with her horrible 'Eulu! eulu!'

At my back I could feel the heightened hatred of this creature who had worshipped her mistress, the woman whose death I had caused. I was afraid to turn round and face her, question her, receive proof of what I already knew. Like a drunken man I staggered on ...

III

When I regained my faculties, I found I was leaning against a marble fountain near the house painted with tulips and yellow butterflies, the house where Aziyade had lived; I was sitting and my head was spinning; the gloomy, deserted houses performed their macabre dance before my eyes. My forehead was banging against the marble of the fountain; it was bleeding; an old black hand, wet with cold water, supported my head ... Then I saw old Kadidja beside me weeping; I gripped her wrinkled, monkey-like hands ... she continued pouring water on my forehead ...

Some men went by but paid no attention to us; they were talking animatedly and reading the hand bills that were being given out in the streets, carrying news of the first battle of Kars. These were the dark days in the early stages of the war, and the fate of Islam appeared to be already sealed.

IV

'I keep awake, and, night and day, my brow, fraught with dreams,
Sends tears streaming down my cheeks
Since Albayde closed her gazelle eyes
In Death's own chamber.'

(VICTOR HUGO, 'Orientales'.)

The cold object that I held tightly in my arms was a marble stone set in the ground.

The marble had been painted azure blue and had at the top a relief of golden flowers. I can still see the raised, gilded lettering that I read automatically ...

It was one of those tumulary stones which in Turkey are specially for   women, and I was seated on the ground in the great cemetery of

Kassim-Pacha.

The freshly turned red earth formed a mound the length of a human body; tiny plants,uprooted by the spade had been placed, roots in the air, on this unplanted spot; moss, herbs, and fragrant wild flowers were all around.--No flowers or wreaths were ever placed on a Turkish grave.

This cemetery had none of the horror that we in Europe associate with our burial grounds; oriental sadness is more gentle, and at the same more majestic. Great stretches of mournful solitude, barren hills, planted here and there with dark cypresses; at intervals and shaded by these huge trees, were lumps of earth turned over the previous day, ancient markers, and curious Turkish tombs surmounted by fezes and turbans.

Below me, way in the far distance, the Golden Horn, the familiar outline of Stamboul, and, over there ... Eyoub!

On that particular summer evening, the ground, the dry grass, all felt warm except the piece of marble that I'd clasped in my arms: it remained cold; its base was sunk in the ground where its contact with death kept it colder still.

Objects round about looked different, as they do when

the fates of men or empires are nearing those mightily decisive turning points that will result in destinies coming to an end.

In the far distance I could hear the fanfares of the troops setting off for the Holy War, those strange Turkish fanfares, each a unison that is both strident and sonorous, with a timbre unknown in our European brass sections; it could have been the supreme 'hallali' of Islamism and the Orient, the death signal of the mighty race of Ghenghis.

I wore the uniform of 'yuzbachi' with my Turkish 'yataghan' at my side. I no longer called myself Loti, but Arif, Yuzbachi Arif-Ussam;--I'd begged them to send me to the front; I was to leave the following day ...

An immense meditative sadness hung over this land that Islam held sacred. The setting sun gilded the old, green-stained, marble tombstones; it moved its pink rays over the tall cypresses, over their ancient trunks, over their grey, melancholy branches. This cemetery was like one gigantic temple to Allah, possessing, as it did, some of the same mysterious calm as well as moving people to prayer.

It was as if I were looking through a funeral veil, and my past life

whirled around in my head as sketchily and disorderly as dreams. I saw every part of the world in which I had lived and loved. I saw my brother, and the women of different colours whom I'd loved. And then, alas, I saw my bloved home that I had deserted forever, the shade of our plane trees, and my old mother ...

For her who lies here, I have forsaken everything!... She loved me with the deepest, the purest, as well as the humblest love: behind the gilded bars of the harem, slowly and gently, sending me no reproach, she died of grief. I can still hear her telling me in that deep voice of hers: 'Me, all I am is a little Circassian slave ... But, Loti, you know that; go if you wish; be guided by what you really want to do.'

Fanfares burst out in the distance; it could have been the trump of doom, as in the Bible, sounding on the Day of Judgement. In one voice, thousands of men shouted out the terrible name of Allah; echoes of the distant clamour reached where I was and filled the vast cemeteries with strange murmurings.

The sun sank behind the holy hill of Eyoub, and the clear summer night descended on the heritage of Othman ...

I shuddered at the thought of that sinister thing lying so close beneath me, already consumed by the earth, that thing I loved still ...

In God's name, is this all there is to it?... Or has some sort of vague entity, a soul, been left behind, hovering here in the pure evening

air, something that can see me, here on earth, still shedding my tears?...

Good God, I'm close to saying a prayer for her; at this moment, this heart of mine that has grown hard and impassive in the course of life's comedy, lies open to all the delightful errors of man's religions, and my tears fall without bitterness on the bare earth. If everything doesn't come to an end in dusty death, I'll know it soon, perhaps. I'm going to try to die, so that I'll know ...

V

CONCLUSION

'Djeridei-Havadis', the Stamboul newspaper, reported:

'Among the bodies of those killed in the last battle of Kars, was that of a young officer of the British Navy who recently enlisted in the Turkish forces under the name of Arif-Ussam-effendi.

'The body has been interred amongst the brave defenders of Islam (may Mahomet protect them) at the foot of Kiziltepe, in the plains of Karadjemir.

F I N

دَوْلَتِ عَلِيّهُ عُثْمَانِيّه

Printed in Great Britain
by Amazon